Y0-BQS-264

DISCOVERERS OF THE LOST WORLD

DISCOVERERS

OF THE LOST WORLD

An account of some of those who brought back to life South American mammals long buried in the abyss of time

GEORGE GAYLORD SIMPSON

Yale University Press: New Haven and London

Designed by Sally Harris
and set in Bembo type by Eastern Typesetting Company.
Printed in the United States of America.

Library of Congress Cataloging in Publication Data

Simpson, George Gaylord, 1902–
Discoverers of the lost world.

Includes index.
1. Mammals, Fossil. 2. Paleontology—South America.
3. Paleontologists—Biography. I. Title.
QE881.S482 1984 569'.098 84-2243
ISBN 0-300-03188-2

The paper in this book meets the guidelines for permanence
and durability of the Committee on Production Guidelines
for Book Longevity of the Council on Library Resources.

10 9 8 7 6 5 4 3 2 1

Contents

Preface

This book was written after I retired from the faculty of the University of Arizona. In addition to my own memories from Scott onward and to other materials for which thanks are given in this preface, the substance is largely derived from the library of the Simroe Foundation.

As ever, the manuscript was read and some corrections made by my wife, Anne Roe. Most of the manuscript was typed by Ronilyn McDonald, and typing was finished by Jon Marks, both transcribing from my handwriting. (My word processor is the connection between my brain and a desk pen.)

As with several previous books, I am indebted to Yale University Press and most particularly to Ellen Graham as editor.

Most of the printed sources for the accounts in this book are cited or fairly evident in the texts of its chapters and the notes following them. I wish here to give thanks also to a number of people who have provided materials not so explicitly noted elsewhere. They are here in the order of the subject matter:

Osvaldo Reig gave me the book on which the interlude about Muñiz is based.

Donald E. Russell sent me copies of unpublished biographical material on Gaudry.

Mary Dawson and Betty Hill sent me copies of unpublished letters and other material by and about Hatcher.

Donald Baird also provided copies of material related to Hatcher and especially also to Scott and to Sinclair. He also arranged a loan of a copy of Hatcher's printed volume on his work in Patagonia.

Larry Marshall gave me copies of mostly unpublished journals by Riggs about his expeditions to South America.

Guiomar Vucetich and Rosendo Pascual had some materials sent from the University and Museum of La Plata.

William A. Clemens provided copies of most of Stirton's field notes and journals bearing on his work in Colombia.

Emiliano Aguirre added material to that available on Botet and the Boscás, helpful for Interlude II.

Zilah B. de Paula Couto filled in details about her late husband Carlos.

Larry Marshall, already named above, also gave me a copy of a manuscript account of the Cretaceous mammal found in Bolivia.

José Bonaparte gave me a copy of a manuscript account of the Cretaceous mammal found in Argentina.

Much of the source material for this book derives from languages other than English. In quoting from letters, papers, or published works, and generally in citing titles of publications, I have used my own English translations. The instances of this—i.e., where the original was in another language—will be evident, I trust.

I have had frequent occasion to refer to museums of natural history in the Americas and Europe. After giving the full formal title in the appropriate language on first mention, I have generally used English (e.g., "the Paris natural history museum" or "the La Plata museum") for subsequent references.

Prelude

"South America is a place I love, and I think, if you take it right through from Darien to Fuego, it's the grandest, richest, most wonderful bit of earth upon this planet. . . . Why shouldn't somethin' new and wonderful lie in such a country? And why shouldn't we be the men to find it out?"

Those are words ascribed to the fictional Professor George Edward Challenger by his creator Sir Arthur Conan Doyle. The thought, although not in those words, had certainly occurred long before and is still occurring to many people, some themselves South American, others in distant countries and continents who nevertheless are attracted to South America and its wonders.

Conan Doyle is best remembered for his detective stories starring Sherlock Holmes, but he considered those mere potboilers inferior to many of his other works, notably his apologia for Britain in the Boer war and his six-volume history of the first (the Kaiser's) world war. In his late years, although he lived only to the age of seventy-one, he was obsessed by spiritualism.

His novel *The Lost World* was an adventure into science fiction, first published in 1912 and original for its time. Although outside the line later followed by this genre, it was a success, provided a basis for an early (quite modified) motion picture, and is still in print and rather widely read. Its plot involves the adventurous exploration of the top of a great mesa in South America where at various times animals and plants became isolated above the high, vertical, impassable walls. It fancies the existence of dinosaurs surviving with little modification from as long past as the Jurassic period, 150,000,000 years or so ago, but it also is a potpourri with imaginary "missing links" and quite modern Indians mixed in with the dinosaurs and other animals. Among nonhuman mammals it specifies, for instance, tapirs, which are alive and well today but are late arrivals in South America, and toxodonts which belonged to an extinct family of mammals of South American origin but are in the novel ridiculously characterized as "a giant ten-foot guinea pig with projecting chisel teeth"—a depiction which toxodonts did not even remotely resemble.

There are a number of such high plateaus with difficult access in South America, notably in southeastern Venezuela and closely adjacent regions, with the (Carib family) Indian name of *tepuis*. Roraima, one of these just on the border of Venezuela and Guyana, has been seriously, or at least semi-seriously, considered as the original of Conan Doyle's "lost world," although Conan Doyle did not locate his entirely imaginary "lost world" in that region and although he gave it the ridiculous name of "Maple White Land" and described its geology in some detail as radically unlike that of Roraima or any other tepui. The tepuis have now been carefully explored, notably by the William H. Phelps (father and son), without discovering any prehistoric animals, although some birds do show slight differences probably evolved in isolation on the tepuis.

So much for the fictional "lost world," an expression which is incorporated in the title of the present book. It is apt here because South America does indeed have a lost world. That world is not living on tepuis (or elsewhere). It is present in the vast extent, both in space and in time, of the geological strata that have been laid down over the hundreds of millions of years of geological time. It is, or rather one should say they are, the many lost worlds of South American life changing constantly and thus becoming lost by extinction, by replacement, and by other changes as time went on.

This book will be mainly concerned, as I myself have often been, with the lost worlds of mammals, the warm-blooded, milk-giving animals like us, although our species is among the very last to appear in South America and thus not long involved in its lost worlds. I have already written a history of those lost worlds as such: *Splendid Isolation: The Curious History of South American Mammals* (Yale University Press, 1980). The present book is not a revision or a repetition of that history. A British reviewer, Gordon B. Corbet, ended his generally favorable discussion of that book by writing, "There are intermittent glimpses of the fossil sites and of the personalities of the palaeontologists who explored them, but in a subject where interpretation has been a distinctly subjective matter one might have hoped for a little more flesh to clothe the bones of the palaeontologists if not of the fossil mammals themselves." I have taken this suggestion seriously.

The present book, then, is not a history of South American mammals. It is a story about who discovered them, who studied them, and to some extent, at least, what the people who thus restored these lost worlds were themselves like. Of each of them we may say, echoing one of Conan Doyle's fictional characters: "You are a Columbus who has discovered a lost world."

Early Days

About sixty miles west of Buenos Aires there is an old town named Luján (pronounced *loo-chan′*, with the *ch* approximately as in Scottish "loch" or German "ach"). It is not large or especially attractive, but almost all Argentinians know it by name at least. It is to Argentina somewhat as Guadalupe is to Mexico and Lourdes is to France, and for a similar reason. Far back in colonial days an apparition of the Virgin Mary was reported here. Ever since then in increasing numbers the pious have made pilgrimages to Luján to ask for help in their afflictions and to give thanks if help is forthcoming.

Some paleontologists do, and all should, know of Luján for two reasons not directly involving piety, which is not necessarily a motivation for paleontologists. Thomas Jefferson probably never knew the name "Luján." There is nevertheless a long, rather complex and tenuous connection between Luján and the occupant of Monticello and of the White House: our third president. I will take up this thread here, trace it back to Luján, and then go forward from Luján to the establishment of mammalian paleontology as a science.

The sequence that we are first following began early in 1796 or possibly the year before and not yet with Jefferson but with workmen whose names apparently were not recorded. These workmen were engaged in digging out saltpeter from a cave belonging to one Frederic Cromer in Greenbriar County, then in Virginia but later included in the severed state of West Virginia. Saltpeter, often found in bat caves, is a nitrate salt with some uses in industry and agriculture but at that time most commonly used in the manufacture of gunpowder. The diggers came across some peculiar bones, which Cromer passed on to a neighbor, John Stuart, who was a friend of Thomas Jefferson. Stuart knew that Jefferson, who was interested in practically everything, would be interested in these oddities, and both he and a relative of his, Archibald Stuart, wrote to Jefferson about them.

Referring to Jefferson's *Notes on the State of Virginia,* which by 1786 had circulated in their first form, John Stuart wrote to Jefferson on 11 April 1796, in part: " . . . observing by your Notes your very curious desire for Examining into the antiquity of our country, I thought the bones of a Tremendious Animal of the Clawed kind . . . might afford you some amusement. . . . I do not remember to have seen any account in the History of our Country, or any other, of such an animal, which was probably of the Lion kind." The parts then in hand consisted of the two bones of the front leg between elbow and wrist (the radius and ulna), three claws of different sizes, and six other toe (or hand) bones. Through Colonel Stuart (whose name Jefferson customarily misspelled "Stewart") and others, Jefferson sought eagerly to obtain more bones of this animal, especially a rumored femur (thigh or upper bone of the hind leg), but these efforts were unsuccessful.

On 3 July 1796 Jefferson wrote to David Rittenhouse, as president of the American Philosophical Society, announcing this discovery. He stated that the bones were those "of an animal of the family of the lion, tyger, panther &c but as preeminent over the lion in size as the Mammoth is over the elephant." The bulk of the claw, he wrote, "entitles it to give to our animal the name of the Great-claw or Megalonyx." Somewhat parenthetically it may be added that "megalonyx," a compound of two Greek words meaning "great claw," was used as an addition to the vernacular by Jefferson and did not thereby become the technical scientific name of a genus of animals. As a technical generic name it dates from Richard Harlan in 1831. To others this sort of thing may be a quibbling detail, but for professional zoologists such niceties are necessary to keep their nomenclature in order.

Rittenhouse had died shortly before Jefferson wrote to him, and the letter was received by Rittenhouse's nephew Benjamin Smith Barton as editor for the American Philosophical Society, then as now the oldest and most prestigious of American learned societies. The year 1796 thus became the election year not only for the presidency of the United States but also for that of the American Philosophical Society. Jefferson was elected to both, and he valued the latter election more highly, calling it "the most flattering incident of my life, and that to which I am most sensible." Nevertheless, his imminent departure for inauguration in Washington made him hasten to write an announcement of the discovery of megalonyx to be read to, and published by, the Society. (It was read to the Society on 10 March 1797 but was not published in the Society's *Transactions* until 1799). When Jefferson wrote the first draft for this announcement he still designated the discovery as "of the family of the lion, tyger, panther &c" as he had in his letter addressed to Rittenhouse and on later occasions. However, after writing that draft and before it was read to the Society and eventually published Jefferson found occasion to have doubts, if not quite to be brought up short.

There came into Jefferson's hands a copy of the September issue of *The Monthly Magazine, and British Register for 1796.* In this was an English version of a memoir by the young French zoologist Georges Cuvier on a recent fossil discovery in South America. Included was a poor copy of Cuvier's somewhat imperfect illustration of a megatherium (Greek "big beast"), which the British magazine captioned as "The SKELETON of a large species of QUADRUPED hitherto unknown *lately discovered one hundred feet under ground near the River la Plata.*" Poor as it was, the illustration showed that this South American creature had claws somewhat like those of Jefferson's North American megalonyx and also that whatever the South American animal might be it was not "of the family of the lion [etc.]"

Although he was apparently still not persuaded, this remote discovery did raise some doubt in Jefferson's mind. He modified the title of his manuscript to "On the Discovery of certain bones of an animal of the clawed kind in the Western parts of Virginia," and he changed the text correspondingly. "Clawed kind" does include lions, tigers, panthers, and a great variety of other animals, and Jefferson continued to compare the megalonyx bones with those of a lion. He may thus fairly be said to have hedged his bet: he evidently still felt that megalonyx was an oversized member of the cat family, but he left open the possibility that he was mistaken.

With the possible exception of Herbert Hoover, whose presidency was ill-fated, Thomas Jefferson was no doubt the most versatile and erudite president the United States has ever had. As a president he was notably well-fated, but like most versatile and erudite people he did have his quirks. His cautiously modified but not abandoned ideas about megalonyx were affected by two of those quirks.

One of Jefferson's opinions, never changed as far as I know, was that no animal that ever lived has become extinct. As late as 1825 (he died in 1826) Jefferson published his conviction that, "Such is the economy of nature, that no instance can be produced of her having permitted any one race of her animals to become extinct; of her having formed any link in her great work so weak as to be broken." Yet as early as 1768 (published in 1769) William Hunter had expressed to the Royal Society in London his opinion about what was then called (among other things) "the American *incognitum*" and is now known as the American mastodon. He concluded by saying that, "though we may as philosophers regret it, as men we cannot but thank Heaven that its whole generation is probably extinct." More directly to Jefferson's point, as early as 1812 Cuvier had published evidence that not only one but many of the "links" in "the economy of nature" (as Jefferson called them) had become extinct and that both Jefferson's megalonyx and Cuvier's megatherium were among them. The eager search for a live megalonyx, and even earlier for a live "American *incognitum,*" was never to succeed.

The other quirk involved in Jefferson's later qualified but never quite abandoned opinion about megalonyx was a sort of running battle that he carried on with the great French naturalist Georges Buffon, whose life (1707–1788) overlapped Jefferson's (1743–1826) for about forty-five years. With unmitigated Old World conceit Buffon had expressed the view that animals of the New World were on the whole smaller than their Old World congeners or equivalents. Jefferson reacted forcefully in his famous *Notes on the State of Virginia*. Those "Notes" were first written in 1781, later revised, printed privately in 1784 and publicly in 1787. Since then they have been variously reprinted. Even in 1781 the fossil animals now called American mastodons (technically *Mammut americanum*) were well known, although still confused with the quite different Siberian fossil mammoths. Jefferson used them as an example of the great size of New World mammals. It is only incidental here, but interesting, that after Meriwether Lewis and William Clark returned from their great expedition to the Pacific coast President Jefferson sent Clark to collect at Big Bone Lick, the best-known locality for remains of fossil mastodons. The large resulting collection for some time occupied a special room in the White House.

In view of the background of Jefferson's disagreement with Buffon, who had in fact been dead for about eight years when megalonyx turned up, it seemed at first irresistible for Jefferson to welcome the discovery in America of a lion-like creature enormously larger than any Old World felines. However, doubt crept in, and in the end Jefferson's comments on megalonyx became a reasonable compromise. "We are to conclude," he wrote, "that [Nature, personified] has formed some things large, and some things small on both sides of the earth for reasons which she has not enabled us to penetrate, and that we ought not to shut our eyes upon one half of her facts and build systems on the other half." Thus this great man simultaneously reprimanded himself and his *bête noire* Buffon.

After Jefferson had finished his revised announcement of the discovery of megalonyx, the specimens were turned over to Caspar Wistar, another distinguished member of the American Philosophical Society and in due course also president of that organization. His work, with well-nigh perfect illustrations, was published in the same (1799) number of the Society's *Transactions* as Jefferson's emended notice. Wistar's conclusions, although cautious and somewhat tentative, were as near perfection as could have been accomplished in the state of knowledge of that time. He recognized the resemblances between megalonyx and megatherium and the likeness of both to living sloths, but considered megalonyx distinct from those animals, living and fossil. Time proved that opinion to be absolutely correct as far as it went, and with time it has gone much farther.

What was then known about megatherium was based almost entirely on a specimen found at Luján. Thus there was a link between Jefferson and Luján via megalonyx and megatherium, and a first link between the animals of lost worlds in North and in South America. Much later, long after Jefferson and Cuvier were dead, the link between megalonyx and megatherium was found to be still more significant and in a way that could not even be glimpsed in the times of those great men. We will turn to that later. Just now it is of special interest to examine more closely that part of the sequence that led from Luján to Paris and to Cuvier.

The megatherium from Luján was not the first fossil mammal found by the Spanish invaders of the New World, but it became the first to be recognized as what it was and to be described and classified in a scientific way. As far as I know the first record of New World fossils was written by Bernal Diaz del Castillo, a former captain of the army of Cortez that conquered Mexico. In his *Historia verdadera de la conquista de la nueva España* (*True history of the conquest of New Spain*), Diaz related that when the army was in Tlascala, Mexico, in 1519 they saw some enormous bones. The natives said that in former times the region had been inhabited by people of great stature and wicked manners who had eventually been extirpated by the ancestral Tlascalans. In evidence thereof they brought a bone which according to Diaz was as long as he was high (certainly with some exaggeration as Diaz called himself "of average size"). The Spanish took this as certain demonstration of the former existence of giants and sent it off to the king in Spain. In fact it was a femur of a mastodont, an extinct animal unknown to be such at that time.

The earliest record of a South American fossil mammal that I know of is of much the same kind. It was made by a Jesuit, José de Acosta (sometimes written d'Acosta). He was in Peru from 1571 to 1587. In 1590 he published an extensive work on "las Indias" (the Indies, as Spanish America was then known) which included among other things the natural history of Peru. In this the presence in Peru of immense bones of prehistoric giants was recorded. These were almost certainly bones of mastodonts.

Such discoveries continued to be made not only in Peru but eventually over much of South America. The earliest of these were summed up by Garcilaso de la Vega, known as "el Inca," who was born in or about 1539 and long lived in Lima. Eventually he moved to Córdoba in Spain, where he wrote and had published in 1609 a *Historia general del Perú*. Included in that work was what Cuvier called a "Gigantologie espagnole" ("Spanish giantology"). "El Inca" died in 1616, doubtless still convinced that there were giants (in Peru) in those days—"those days" being before the coming of the Incas. As will appear later in this account, the real nature of those giants was pretty well cleared up by Cuvier some two centuries later. We will come to that in due course, but first let us go back to Luján and the megatherium.

The discovery at Luján and its subsequent history are largely documented, although not in really complete detail. Two errors can be cleared out of the way at once. The statement in *The Monthly Magazine* that the skeleton was discovered one hundred feet underground is obviously untrue, for the region where it was found was only some thirty to thirty-five feet above "water level," and no cut bank, excavation, or well can have been made there to a depth of a hundred feet in the latter part of the eighteenth century. Since the skeleton is nearly complete and of one individual, it must have been buried whole in sediment at death, or shortly after, and it cannot have been fully exposed by recent stream erosion. It was said to have been found by a Dominican Friar, Manuel Torres, on the banks of a small tributary to the Luján River. Who in fact first saw it and who dug it out are not matters of known record.

Another known error, variously copied, gave Paraguay as the geographic source of this specimen. There was a specimen of megatherium found in Paraguay and sent to Spain, but this was well after the first specimen reached Spain and it was less complete than the first. According to Cuvier's subsequent account this partial megathere was given by a lady (unnamed) as a present to one Father Fernando-Scio of the "Ecoles Pies" (parochial schools—*Pies* can be read either as "pious" or as "charitable"), presumably in Madrid. That was the third specimen known to Cuvier in 1812, a second, also inferior, one having been sent in 1795 from Lima (in Peru) to the same royal collection in Madrid as the first and most complete specimen.

There is some doubt as to when the first and most complete megatherium was found near Luján and when it arrived in Madrid. It was probably found in 1788 and reached Madrid, packed in seven boxes, either later that year or in 1789. By 1789, at latest, it was placed in the Real Gabinete de Historia Natural (Royal Cabinet, in this usage essentially a museum, of Natural History), which had been founded in 1771 by King Charles III. Here it was placed in the hands of Juan Bautista Brú y Ramón, who had been employed in the museum since its foundation. Brú (sometimes written without the accent, which does seem superfluous) was a sort of man of all work, more or less what is called a preparator in modern museums but also with some pretensions as a scientist. In 1784 he had published the first volume of a work on the collection of "animals and monsters" in the museum. Among its many illustrations was an engraving of the skeleton of an Indian elephant which Brú had assembled and mounted in a lifelike pose. The book as a whole was a hodgepodge and was not a systematic contribution to science, but the mounted skeleton of the elephant was a prelude to what was to give Brú an honorable, if somewhat secondary, place in the history of science.

Having unpacked the megatherium bones, Brú cleaned them off and then assembled and mounted them in a more or less lifelike pose. This was the first fossil skeleton ever to be mounted in such a way, and that is Brú's well-deserved right to fame—well-deserved even though the later discovery of many megatheriums and their more intensive study and comparison indicate that Brú's original mount was awkward and not very lifelike. This is seen in a drawing of the mounted skeleton, made under Brú's direction and under circumstances to be discussed later in this chapter, first published by Cuvier. The skeleton is still displayed in the Museo Nacional de Ciencias Naturales in Madrid, but its pose has been somewhat changed from that given it by Brú, although still not wholly lifelike.

Regarding Brú as a person, rather little is known and that little is not flattering. The date of his birth is unknown but he was adult in 1771. It is known that he died in December 1799. He seems to have been temperamental and difficult to get along with. In 1794, after more or less twenty-three years of work in the Gabinete, he went over the director's head and addressed a minister of the royal court asking to be made deputy director of the Gabinete. The director not unnaturally found this action unjustified and its pretension absurd. Thereupon Brú impetuously left the Gabinete, and he took with him his drawings, the engravings of them, and his description of the megatherium skeleton.

Brú certainly knew that he had in his hands one of the greatest discoveries ever yet made and that priority of its publication would take him out of relative obscurity to absolute fame among scientists. This did not happen, for reasons which remain somewhat obscure but which may have involved what I would call in my western vernacular some dirty work at the cross-roads. In 1795 one Philippe-Rose Roume, an official of French West Indian Saint-Domingue (or Santo Domingo, just ceded to France in 1795; now the Dominican Republic), was traveling from the island to France by way of Spain. In Madrid he "chanced upon" or "had occasion to obtain" proofs of the engravings that Brú had had made for his planned publication on the great new fossil from Luján. Roume was a member of the Institut de France, then recently founded, of which Cuvier had also recently become a member. Roume sent these engravings to the Institut, according to Cuvier without descriptions and only with a short note of sorts ("une courte notice de sa façon"). The section (*classe*) of sciences of the Institut referred Roume's loot to Cuvier, already well known as a student of such things although he was only twenty-six years old.

The young Cuvier, no less than the aging Brú, knew a good thing when he chanced upon it. He forthwith wrote the first of what would be many publications on fossil mammals and other fossil vertebrates. His title, here translated, was "Notice on the skeleton of a very large species of quadruped hitherto unknown, found in Paraguay and deposited in the natural history collection [*cabinet*] of Madrid." This was published in the *Magasin encyclopédique; ou; journal des sciences, des lettres et des arts* of Paris in 1796. It included, as Cuvier wrote later, "a bad copy of the figure of the whole skeleton." This was the article hitherto mentioned that became known to Jefferson as summarized in English in *The Monthly Magazine* with an even worse copy of the bad figure. As Cuvier later explained, this was about the skeleton from Luján (Cuvier used the alternative spelling "Luxan"), and its attribution to Paraguay was a mistake.

Thus Cuvier, who never had seen and never did see the actual bones of the animal he described and named, obtained priority of publication of this great discovery. Meanwhile, poor Brú back in Madrid was getting into still more trouble taking him further out of the limelight he deserved. One José Garriga came across Cuvier's publication, noted that the skeleton described was in Madrid, not Paris, and decided to defend the priority of Spain and of Spanish science. Somehow Garriga induced Brú to sell him the drawings, plates, and description of the giant fossil. He then, still in 1796, published in Madrid a separate folio with the title (here translated) "Description of the skeleton of a very corpulent and rare quadruped which is preserved in the Royal Cabinet of Natural History in Madrid." Although he said that he was doing this to obtain credit for Brú and to correct errors by Cuvier, he derived the substance from both Brú and Cuvier. If one cannot help thinking that Cuvier had not been completely ethical in this affair, one also cannot help thinking that Garriga's inducing Brú to turn everything over to him and making this the basis of his own publication was not fully ethical either, even though he gave credit to Brú.

Baron Georges Léopold Chrétien Frédéric Dagobert Cuvier, to give him his final, full, and resounding name, became one of the greatest scientists of his time. It is a peculiar fact that the boy who eventually amassed that title and the following assembly of Christian names was originally christened as Jean-Léopold-Nicolas-Frédéric, but the Jean was dropped and Georges added. (I don't know when or why Nicolas was replaced by Chrétien and Dagobert was tacked on at the end.) It is not surprising that he has always been known simply as Georges Cuvier. His younger brother, christened Georges Frédéric, lost the Georges and was known simply as Frédéric.

The Cuvier was born on 23 August 1769 in the small town of Montbéliard, French-speaking and Lutheran but then attached to the German Duchy of Württemberg. (He did not legally become a French citizen until 1793.) As a child he was a brilliant student, and the Duke of Württemberg entered him in the Karlsschule established in Stuttgart by the duke for the most talented youths in his duchy. Between primary schools and the Karlsschule the still young Cuvier is recorded as having studied and mastered civil history, religious history, Latin, Greek, geography, mathematics, law, chemistry, mineralogy, zoology, botany, mining, police work, diplomacy, commerce, finance, economics, and geometry. Apart from all that he also had private instruction from Karl Friedrich Kielmeyer (1765–1844), a naturalist who was a forerunner on one hand of evolutionism of a sort and on the other hand of the German school of "Naturphilosophie." It was more this personal instruction than all the formalized pedagogy that turned Cuvier into the foremost anatomist, taxonomist, and zoologist in general of his time, and also into the "father of paleontology" as a science in its own right. It is the latter role in its connection with South America that most concerns us here.

The first notice or memoir on the megatherium in 1796, as already noted, was Cuvier's first publication on a fossil of any sort, and it happened to be a South American fossil mammal. In most years thereafter he published from one to nine papers, memoirs, or books about various fossils, but relatively few of those referred to South America. For many of his materials he had to go no farther than to the Butte de Montmartre, then being quarried for gypsum, in which many fossil animals were entombed, now completely built over by the expansion of Paris and the grandiose but tasteless basilica (started in 1875). Paleontologists would like to see all the buildings razed and Cuvier's fossil beds reexposed, but that is hardly practicable.

In 1804 Cuvier published another, somewhat modified notice "On the *Megatherium,* another animal of the sloth family but of the size of the rhinoceros, an almost complete skeleton of which is conserved in the royal cabinet of natural history in Madrid." His definitive discussion of this skeleton, which he had never seen and never did see, was included in the work that has led paleontologists to consider him the principal founder of their science: "Researches on the fossil bones of quadrupeds in which are established the characteristics of various species of animals that the revolutions of the globe seem to have destroyed."

This work is now most commonly called the *Ossemens Fossiles* even by English-speaking students. Its first edition was published in four volumes in 1812. For the most part it consists of a compilation of studies previously published separately and here collected with such revisions and additions as Cuvier could make. In Volume IV Cuvier preceded discussion of the megalonix (as he spelled it) and the megatherium with descriptions, measurements, and figures of the skeletons of the two genera of living tree sloths, the *ai* (three-toed sloths; in present classification the genus *Bradypus*), and the *unau* (two-toed sloths, genus *Choloepus*). His purpose for this was to make comparisons of these South American animals with the likewise South American "megatherium" and the North American "megalonix." Treating megalonix first, he cited Jefferson and the hint, which remained in Jefferson's revised text, that this still might prove to be an enormous, extant member of the cat family. It is odd that Cuvier did not mention the contrary opinion of Wistar, which as I have already noted appeared in the same publication as Jefferson's account.

Although Cuvier still had not seen the megalonix bones, and never did see them, he now had the best substitute. Charles Willson Peale, the eminent artist best known for portraits of George Washington, was resident in Philadelphia where the bones then were, and he made plaster casts and sent them to Cuvier. As Cuvier wrote (in French), Peale "thus has given me the ability to describe them all anew and to have illustrations made from points of view somewhat different from those of M. Jefferson." Cuvier's seventeen figures of the casts, on a single large plate, are more varied than, but inferior to, those already published by Wistar.

There had been an apparently bitter dispute about megalonix and megatherium in Paris between Cuvier and Barthélemy Faujas de Saint-Fond, well known as a geologist and an expert on volcanoes but not on fossil bones. Addressing him with ill-disguised irony, Cuvier wrote, "M. Faujas, my learned colleague in the Museum of natural history, has carried over the name of *megalonix* to a fossil animal of a different species, although of the same family, discovered in Paraguay, which he has not distinguished from that of Virginia, although they are quite different, as we shall see." In other words Faujas had had the nerve to extend Jefferson's name megalonyx, or megalonix, according to the French writers, to Cuvier's megatherium. Faujas had even gone so far as to adopt the view that Jefferson at first espoused but then only skirted: that megalonix was an enormous, ravening carnivore.

Cuvier demolished Faujas's views and those of some other disputers. He showed beyond any possible doubt that megalonix and megatherium were quite distinct from each other, although related, and also that both were distinct from the living tree sloths but belonged to the same "family." Now a mature scientist in his early forties and known world-wide, he carried the day, and his views on these points were soon acknowledged to be essentially correct—as they still are. In this connection, however, it is to be remembered that in Cuvier's day the technical category of "family" in classification was much broader than it is now. At present two distinct families, Megalonychidae and Megatheriidae, are recognized, each now with many known extinct relatives, and they are considered related to each other and also less closely to the living sloths at the level of superfamilies or the still broader category of an infraorder.

The definitive treatment of the megatherium by Cuvier follows that of the megalonix in the first edition of the *Ossemens Fossiles*. Cuvier's own contribution on megatherium in this work summarizes the history of the then known three specimens of megatherium, all in Madrid and none actually seen by Cuvier. He then compared megatherium and megalonix with each other and with living sloths and their likewise South American relatives. The conclusions, here translated from Cuvier's French, were:

> As for the comparison between the *megatherium* and the *megalonix*, it gives as result an almost complete identity in forms [descriptive anatomy] at least in the parts known in the latter, but the size is different. The bones of the *megatherium* are a third larger than those

> of the *megalonix*, and as the latter have all the characters of adult status, one can hardly attribute this difference in size to anything but a difference in species. One may add that the claws have their sheaths more complete and longer in the last phalanges of the *megatherium* than in those of the *megalonix*. These two animals will then have formed two species of the same genus, belonging to the family of the edentates and serving as intermediates between the sloths and the anteaters, closer, however, to the former than to the latter.

Cuvier added that it is remarkable that the fossils and their closest living relatives were found only in "America," which for him included North, Central, and South America.

To that discussion Cuvier appended an abridged but still lengthy description of the megatherium skeleton, translated into French from Brú's original manuscript as published in Spanish by Garriga. This is purely descriptive and does not discuss the relationships of the animal. Cuvier here illustrated megatherium in two plates, mainly reduced reproductions of Brú's figures but with three additional figures by "D. Joseph Ximeno" (that is, Don José Ximeno or Jimeno). Curiously these were given to Cuvier by Faujas, with whom Cuvier was having if not a feud at least a sharp difference of opinion. Just how Faujas got them is not stated, although the artist is named.

In the first edition of the *Ossemens Fossiles* Cuvier devoted the latter part of the second volume to the group we now call "proboscideans" because the living species have trunks (proboscises) as did most if not all the extinct species. As was well known long before Cuvier's time and is now well known to almost everyone, there are just two living genera and species of this group: the Indian elephant (*Elephas maximus*) and the African elephant (*Loxodonta africana*). It had also long been known that a similar animal, called "mammoth" in English and "mamont" in Russian (presumably an adaptation of a Siberian Tatar word) was abundant in fossil remains in Siberia and was also found elsewhere in Eurasia.

A more recent and for a time even more puzzling discovery of an elephant-like fossil animal had been made definitively in 1739 by Charles Le Moyne, second Baron de Longueuil, near the Ohio River in what was then French territory. (It is now in Kentucky.) A large tooth of this creature was first figured in a French publication by Jean Etienne Guettard in 1756. The tooth was elephant-sized, but it was decidedly different from any tooth of a living elephant or of the Siberian mammoth. Guettard's conclusion about it was to ask two questions and not answer them. He wrote (in French), "Of what animal is this [tooth]? And does it resemble the fossil teeth of this size that have been found in different places in Europe? Those are two points that have not been possible for me to clear up."

A partial but rather unsatisfactory answer was achieved by Louis Jean Marie Daubenton (to give him his full name) in a publication of 1764. He showed that bones found in the same deposit as the teeth were closely similar to those of a Siberian mammoth and a living elephant, so much so that he considered all three to be of the same species. As for the teeth, they obviously were not those of mammoths or elephants so in his opinion could not belong to the same animals as the bones. Daubenton decided that they were probably those of a large hippopotamus.

Daubenton's error was soon corrected. An Irish-English trader (and fighter) with the Indians, George Croghan by name, in 1766 collected some bones and teeth at Big Bone Lick, already famous for such discoveries, and in the following year he sent these to London, where they were divided between Lord Shelburne, then in charge of the American colonies, and Benjamin Franklin. Franklin saw that the grinding teeth could not possibly be those of elephants, but he believed that the bones, tusks, and "grinders" (molars) belonged together in a single species of animal. That view was more definitely and publicly supported by Peter Collinson in a paper read to the Royal Society of London in 1767 and printed in the *Philosophical Transactions* in 1768. As to identification Collinson considered these fossils to belong to a distinct species of elephant or some altogether different animal, in either case one not yet known. There is here an echo, at least, of the view that the animal was not extinct but still lurking somewhere in the wilderness. Jefferson of course shared that view as he examined his collection of such remains in the White House.

Most important for the present book is a remark in Franklin's letter of thanks to Croghan. After saying that no such animals were known to be extant, Franklin added, "It is also puzzling that . . . no such remains should be found in any other part of the continent, except in that very distant country, Peru, from whence some grinders of the same kind . . . are now in the museum of the Royal Society."

It is now known that the Proboscidea as a whole have had a very long and very complex history involving all the continents except Australia and Antarctica. By the end of the eighteenth and in the beginning of the nineteenth century, finds of fossil proboscideans were piling up, and they still are today.

In the first edition of the *Ossemens Fossiles* Cuvier went far toward clearing up the confusion that had arisen about the already known late (Pleistocene and Recent) proboscideans. He first divided all of those into two major groups. To one group he gave the vernacular French name "éléphans," simply "elephants" in English. He showed that the Indian and African elephants are sharply distinguishable species as regards their teeth and some other characters. He classified them as *elephas indicus* and *elephas africanus,* clear enough but not valid as technical names under current codes of scientific nomenclature.

Cuvier also showed that the abundant Siberian "mammoths" (*mammouths* in Cuvier's French) are a distinct group of elephants, more like the living

Indian than the living African elephants. In French vernacular he called them simply *"éléphant fossile"* but equated that with the more technical name *elephas primigenius* citing [Johann Friedrich] Blumenb[ach] as its author. Although Cuvier's concept of this group was based primarily on the Siberian specimens, he extended the specific term to include all the most elephant-like fossils found widely over Eurasia and also in North America. Later discoveries and studies have shown that these widespread fossils, all members of the family Elephantidae, nevertheless belonged to a considerable number of different species in the modern sense and also to several groups generally ranked as genera, that is, as differing from each other more than do species and often including more than one species within the generic group.

In this discussion by Cuvier, there is one brief comment that is puzzling. Near the end of the chapter on "éléphans" he remarks, "We have shown moreover that the mountains of the Isthmus of Panama have not been an obstacle to their passage into South America." I cannot find elsewhere in Cuvier's work any evidence that "éléphans," that is, members of our family Elephantidae, did ever cross the Isthmus of Panama, and to this day not a single scrap of that family has been found in South America. It is a curiosity of geographic faunal history that the Elephantidae, although spreading well south in North America, evidently did not reach South America even at the height of faunal interchange between the two continents.

In the *Ossemens Fossiles* all the other proboscideans known to Cuvier were classified as "mastodontes," a term coined by Cuvier in 1806 from the Greek *mastos* (breast) and *odon* (tooth), in reference to the presence of mammillary projections on the low-crowned masticating teeth in contrast with the transverse plates on the high grinding teeth of elephants. The much older term "elephant," in all its linguistic variations, is simply from the classical Greek *elephas,* referring primarily to the Indian elephant and extended to the African elephant and eventually also to the extinct species distinctly related to these.

The part of the *Ossemens Fossiles* devoted to mastodonts starts with the animals first found near the Ohio River and later over much of North America. The discussion thus begins grandiloquently:

> Not only is this the largest of all fossil animals, it is also the first that has convinced naturalists that there might be extinct [*détruites*] species. The monstrous size of its grinding teeth, the formidable projections with which they bristle, cannot indeed fail to draw attention, and it is quite easy to make sure that none of the big animals that we know has had either this form or this bulk.

Cuvier may well be excused if his enthusiasm carried him away a bit in this exordium. It is true, as it is of most mammalian species, that the teeth of this species are unique in detail, and it is also true that this was among the larger land mammals. Nevertheless it was no larger, and indeed overall a bit smaller, than the largest African elephants already known in Cuvier's time. (The later discovery of much larger fossil land mammals, notably the gigantic Asiatic rhinoceros *Baluchitherium,* was of course unknown to Cuvier.)

Because this species was first known from Big Bone Lick near the Ohio River, it was often called by Cuvier and others "the Ohio animal," but it happens that its first virtually complete skeletons, already known to Cuvier but not personally seen by him, were found and excavated by the Peale family in the Hudson valley. Even in Cuvier's time they were known to occur over much of North America, and their geographic range has since been greatly extended. It is, however, still true, as Cuvier noted, that no mastodonts of just this species have been found in South America.

Cuvier was annoyed that many Americans, who evidently had not seen teeth of the extinct Siberian elephants, persisted in calling the Ohio animal a mammoth. In fact through a quirk of the code of technical nomenclature the valid name of the North American mastodon is *Mammut americanum,* as noted above, and that of the Siberian elephant, or true mammoth, is *Mammuthus primigenius.* In the vernacular, "mastodon" is now usually used for the American mastodon, and "mastodont" is usually used for a more varied and widespread group of proboscideans not all closely allied to the American mastodon or to the elephants and the mammoths, strictly speaking.

The third section about proboscideans in the *Ossemens Fossiles* is "On various teeth of the genus of the mastodonts, but of smaller species than those of the Ohio, found in several places of the two continents." The *deux continents* here are not North and South America but Europe and South America. Here Cuvier discusses and compares a number of fragmentary or single and worn teeth and some jaw fragments that are quite distinct from the much better known animal we now call the American mastodon but that are clearly "mastodontes" and not "éléphans" in Cuvier's terms. Most of the teeth in this group are even more mastodont in a descriptive sense, that is, have more nipple-like protuberances or cuspules, than the American mastodon. Cuvier claimed that these smaller mastodonts were "entirely unknown" before his researches. As has already been noted here, some of their remains in South and Central America had in fact been known for a long time, but they had been considered as the remains of human or anthropomorphic prehistoric giants. Cuvier's achievement was showing that they were in fact mastodonts, in his term, and hence resembling and somehow classifiable with the elephants. They were in fact proboscideans, in later terminology for the whole assemblage of trunked or broadly elephantine mammals.

Most of Cuvier's South American specimens in this group were found by Humboldt, who thus deserves special consideration here as the discoverer of South American mastodonts that were classified as such. Friedrich Heinrich Alexander Freiherr [=Baron] von Humboldt, to give him his full name and title, was born in Berlin in 1769 and died there in 1859, not far short of his ninetieth year. In the meantime he had led an extraordinarily varied and adventurous life. Of most interest here were his travels in the Americas in 1799 to 1804, mainly in South America but also in Cuba and Mexico and briefly in the United States. He continued to travel widely but more briefly, and he wrote much in varied fields of science, coming to be considered, "with the exception of Napoleon Bonaparte . . . the most famous man in Europe" (as his biographer Helmut de Terra put it). At the age of seventy-six he undertook to write a full and considered treatment of the whole earth and indeed of the universe under the name of *Kosmos*. Four volumes had been published and he was well into a fifth when he died. The accomplishment was tremendous, but one can fairly say that the cosmos appeared quite a bit younger and less sophisticatedly complex when Humboldt tackled it than it does now.

That Humboldt laid the basis for the recognition of South American mastodonts may seem a small part of such a life, but is a large point in the expanding of the South American lost world of prehistoric life. The specimens brought back from South America to Europe by Humboldt were described and most of them figured by Cuvier. Two were fragments from the end of a molar, both from a locality where many fossil bones were reported. That was near Santa Fé de Bogotá in Colombia at a locality known as what Cuvier called the "Camp-des-Géants"—presumably "Campo de los Gigantes" in Spanish. A nearly complete molar with three worn crests was found near the volcano Imbaburra in the "Kingdom of Quito," now Ecuador. Another, a mere scrap not figured by Cuvier, was from somewhere near Tarija in Bolivia. This is worthy of note because much later this general area was found to be exceptionally rich in fossil mammals of relatively late (Pleistocene) age, and collections are still being made there. Another specimen, figured by Cuvier as a deeply worn, three-ridged grinding tooth, was found near Concepción in Chile.

Cuvier's figure of a fairly complete grinder from roughly the same region around Tarija was based on a drawing sent by "M. [Monsieur] Alonzo," about whom I know nothing more. Three other specimens with no more precise localities except simply Peru were brought back by Joseph Dombey, a relatively insignificant French naturalist about whom there is a brief and not fully informative French biographic sketch. He joined a Spanish expedition to Chile and Peru led by Ruiz and Pavon in 1771 to 1788. What he collected was to be shared with the Spaniards, but he left the expedition early, in 1784, with all his collections and apparently with some idea of publishing them. He claimed that his Spanish companions seized all his drawings (some three hundred of them), that they forced him to swear that he would not publish until the expedition returned, and that they even tried to have him killed. Apparently he never did publish, but he must have turned over some of his specimens to Cuvier, who did publish on them with the comment that they were brought to France by Dombey.

Throughout this study of the South American mastodonts Cuvier was comparing and even mixing them with similar and then equally scanty bits of mastodont teeth from Europe, mostly France. They are indeed similar, and modern classifications agree with the basic features of Cuvier's arrangement, on a great deal more evidence and detail, and also in different terms. Conservative modern classifications recognize the same three main groups of proboscideans that Cuvier did. Cuvier's vernacular term "éléphans" now is the family Elephantidae. His "Ohio animal" and relatives and ancestors unknown to Cuvier are the family Mammutidae. And now the much larger and more complex group called by him "smaller species than those of the Ohio" is the family Gomphotheriidae, including both Cuvier's European "species" and his South American "species."

Among the "lesser mastodonts" known to him, Cuvier recognized four species. According to his custom in 1821 and earlier he named them in vernacular terms although other zoologists and paleontologists were already using the more fixed and technical system of nomenclature formalized in 1758 by Linnaeus. This basically required two neo-Latin names, the first being the name of a genus and the second the name of a single species within that genus. Cuvier called the two European species of the "lesser mastodonts":

> That one of Simorre [a French village] and elsewhere, *mastodont with narrow teeth* [*mastodonte à dents étroites*].
>
> That one with little teeth, *small mastodont* [*petit mastondote*].

The two species from South America were named:

> The big one with square teeth, *mastodont of the Cordilleras* [*mastodonte des Cordilières*].
>
> And the littlest, *Humboldtian mastodont* [*mastodonte humboldien*].

(Cuvier consistently misspelled Humboldt's name, or perhaps Gallicized it, as "Humbold.")

Many different technical names were later applied to the various South American mastodonts, and the situation became extremely confused. It is not at all clear which now-acceptable specific names apply to Cuvier's two claimed species. Mastodonts are known to have been abundant and diverse over practically all of South America. The system I prefer (because it was worked out by the Brazilian paleontologist Carlos de Paula Couto and me) puts all the South American mastodonts in a single subfamily of the Gomphotheriidae and divides them into four genera (one dubious) with only five species in all.

Cuvier had no real misgivings about recognizing such apparently closely similar "species" in South America and in Europe, with none like them known in either Asia or North America. Although it breaks the generally temporal sequence of the present book, a few words about this may best be put in here. Since Cuvier's time, many members of the family Gomphotheriidae, mostly older than those in South America, have been found in the Old World and North America. They spread widely over Africa, Europe, and Asia and from Asia soon reached North America over the Bering land bridge from Siberia to Alaska. As they evolved, they split into a number of different lines of descent. By two or three million years ago these moderately divergent groups were well developed in North America. When the Panama land bridge arose, surely two, probably three, and perhaps even four of these lineages spread to South America and then thrived in appropriate environments there for a million years or more before they became extinct for reasons unknown.

Cuvier did not see any reason to think of an actual connection between groups such as the South American and European mastodonts which he did consider closely related in some nonevolutionary sense. Cuvier's life (1769–1832) widely overlapped that of Lamarck (1744–1829). Cuvier was forty years old in 1809 when Lamarck published the first book ever to advance a flat statement of the evolution of all forms of life. Cuvier was impressed with that work, but in a wholly negative way. He was violently opposed to all ideas of evolution (or of transformation as this was then called).

Cuvier was the greatest anatomist and taxonomist of his time, and close to being as great as any before or since, but his mind was closed on this idea, even as a possibility. One of his reasons was that Lamarckian evolution depended in large part on the very ancient concept of a ladder of nature or a chain of life. This was the idea that the pattern of creation or the nature of life represents a continuous sequence from the simplest or lowest beginnings to the most complex or highest state. Lamarck accepted this as did almost every naturalist before him and many later. (One can see this in Jefferson's remark, previously quoted, that nature would not permit any line in the chain to be broken.) Lamarck's theory largely rested on the idea that the ladder was ascended by evolution. Cuvier rejected the idea of *a* ladder of life because as a comparative anatomist he found not one but several quite distinct lines or branches (*embranchements*) of anatomical patterns. He would have been shocked if he could have foreseen that these observations of his were later found to be strongly indicative of evolution.

Cuvier was one of the principal founders of the most basic historical principle in paleontology: that of faunal succession in which species of ancient faunas became extinct, after which they were replaced by different faunas. In a long (116 page) "Discours Préliminaire" in the *Ossemens Fossiles* in 1821 and in a later re-expression and expansion of this he completely rejected the idea that these faunal successions involved evolution. Instead he insisted that they followed mass extinctions caused by successive but relatively few catastrophic "revolutions of the surface of the globe."

Cuvier became a great man, the leading zoologist of his day and really the first specialized paleontologist of any day. He was apparently modest enough as a young genius, but as his fame grew he became anything but modest. To be sure the title page of his *Ossemens Fossiles* indicates that it is simply by "M. [for Monsieur] Cuvier," but this is followed by twelve long lines in small type recounting the multitude of awards and distinctions that had already been bestowed upon him by 1821, and portraits of him in his prime show him as cold, pompous, and highly decorated. In addition to his science, he was also an adroit administrator and politician: he was eventually grand officer of the Legion of Honor, a peer of France, and president of the Council of State—all this and more as a Lutheran in a Catholic country.

It is difficult to think of Cuvier in his private person. One hardly believes that he had one. I find it almost impossible to clothe him in human flesh, as requested by the reviewer of *Splendid Isolation* whom I quoted a few pages back. Yet he did have some private life: on 2 February 1804, when he was thirty-four years old, he married a lady with the elegant name Anne-Marie Coquet de Trazaille Duvaucel, and in due course they had four children.

Notes

If, as I hope, the reader will want to know more about Cuvier as a scientist, I recommend the following excellent book:

William Coleman. 1964. *Georges Cuvier: Zoologist.* Cambridge: Harvard University Press.

Although Humboldt has only a brief part in this chapter, that may be a reason for a reference to a life of this almost incredibly versatile man:

Helmut de Terra. 1955. *Humboldt: The Life and Times of Alexander Humboldt.* New York: Alfred A. Knopf.

The early history of "the Ohio animal" and of Megalonyx and Megatherium is given in more detail in the following two papers:

G. G. Simpson. 1942. The beginnings of vertebrate paleontology in North America. *Proceedings of the American Philosophical Society,* 86:130–188.

Julian P. Boyd. 1958. The Megalonyx, the Megatherium, and Thomas Jefferson's lapse of memory. *Proceedings of the American Philosophical Society,* 102:420–435.

I would like to know more about Brú, but all I find available is summarized in Boyd's paper, just cited. As for Jefferson, accounts of his life and works are in almost any library. Wistar is less generally known now, but the facts about him really relevant here are summarized in my paper cited above. His dates are 1761–1818, and the wisteria, although misspelled, is named after him. The Wistar Institute is not—it is named for his great-nephew Isaac Wistar.

Darwin

After discovery of the megatheres and the identification of South American mastodonts as such, there was for some time little advance in the collection and study of prehistoric South American mammals. Most important before Charles Darwin, whose South American fossil gathering will be the main topic of this chapter, was Alcide Dessalines d'Orbigny (1802–1857), who was born in France, more exactly in the insignificant village of Couërzon, Loire Inférieur, and who early showed a bent for zoology and paleontology. In his twenties he was appointed as a traveling naturalist for the Paris Muséum d'Histoire Naturelle, and one of his first assignments was to go to South America to make collections for that museum. In the course of his travels he wrote a letter, published in the *Comptes Rendus* of the Académie des Sciences, claiming priority for the discovery of fossil shells and of mastodont bones in the high Andes of Bolivia. As regards the mastodonts, at least, he had long been antedated by Cuvier, as was noted here in Chapter 1. D'Orbigny's best known work, however, was *Voyage dans l'Amérique Méridionale* in several volumes, including one on paleontology. He had discovered fossil mammals in the vicinity of Santa Fé Bajada in northern Argentina, and knowledge of this was a main reason for Darwin's later visit to that locality. D'Orbigny's own interest, however, had been and remained in fossil shells and their geological distribution, especially in France, where in 1853 he became the professor of paleontology in the natural history museum in Paris. He died only four years later.

The other clue to fossil mammals near Santa Fé in northern Argentina known to Darwin was that Hugh Falconer had seen some fossil bones in a stream not far south of Santa Fé. Falconer (1808–1865) was a British doctor who went out to India in 1830. There he discovered many fossil mammals and became famous as the paleontologist who did most to find and describe the very rich Siwalik faunas, now still under intensive study because of the presence of manlike apes in them. He wrote many notes on fossils from several other regions as well, but published nothing on South American fossils. I must confess that I do not know what he was doing so far off the beaten path in that country or how Darwin found out about it.

Almost everyone knows something about Charles Darwin, but some of the things they "know" are not completely true. For example, it is not true that Darwin was the discoverer, or inventor, of "the theory of evolution." In fact various theories of evolution had been proposed before him, including one by his own grandfather, Erasmus Darwin, who died seven years before Darwin was born. The most serious of the forerunners, Lamarck, published his major work on evolution in the very year that Darwin was born (1809). For Darwin evolution became not a theory but a fact, and it has become so accepted by all competent students of the subject. Darwin's theory about how evolution works was that it does so largely, but not solely, by natural selection. With some modification that theory, as distinguished from the fact of evolution, is considered valid by almost all biologists.

Another thing that many people "know," especially if they oppose this idea, is that "Darwinism" means that humans evolved from apes. Darwin never said so. He did say, and again all competent biologists follow him in this, that there were once ancestors in common from which apes, on one hand, and humans, on the other, later evolved in their own separate ways.

Most pertinent in the present context is the fact that Darwin visited South America and that he there collected some fossil mammals, along with other fossils and a great many more things. The extent and the way that his fossil finds contributed to his later growing thoughts about evolution are open to some discussion. That will appear later in this chapter.

Just for rapid orientation, Darwin's life before he set foot in South America should be outlined. He was born on 12 February 1809 (the same day and year as Abraham Lincoln) in the family home in Shrewsbury, the county seat of Shropshire, and he went to the Shrewsbury School, still active today. As his father and grandfather had both been doctors (wealthy ones), Charles was sent off to Edinburgh to study medicine. He found this unbearable, and so as a second best he was next sent to Cambridge with a view toward entering the clergy. He then had no doubts, although no great interest, in the validity and mission on earth of the Church of England. His personal taste turned more and more to natural history, which at that time was a proper and usual hobby for a clergyman.

The common and to most people the only known picture of Darwin is as a rather sad-looking old man with a receding forehead and a tremendous white beard. We sometimes forget that an old man was once a young man. At this earlier period in his life, within a year or two of twenty, Darwin's real passion, as he wrote in his autobiography, was for "shooting and for hunting, and when this failed for riding across country." He got into a "sporting set including some dissipated low-minded young men." With them at dinner he "sometimes drank too much, with jolly singing and playing at cards afterwards." These carousals, which may have been milder than the aged Darwin's nostalgia made them, might have interfered with his religious studies, but not with his greater passion for natural history, collecting and tramping to know the wild living things and the terrane of the world around him.

In 1831, when Darwin was twenty-two, the direction of his life and its eventual outcome were radically changed. A Captain Robert Fitz-Roy was taking a small naval vessel, H.M.S. *Beagle,* to chart the coasts and islands of southern South America and then to continue westward around the world. He proposed to take along a naturalist, without pay. Darwin's application for the post met with some resistance from his own father and initial reluctance from Fitz-Roy. But with support from his uncle (and future father-in-law) Josiah Wedgwood and a recommendation from John Stevens Henslow (a Cambridge professor with whom Darwin took botanical jaunts) he prevailed. On 27 December 1831, Darwin set out on the voyage of the *Beagle,* destined to be made famous by him and to contribute to his own fame. He returned, a greatly changed and more mature man, almost five years later on 2 October 1836.

After some brief previous stops the *Beagle* reached Bahia (also called San Salvador) in Brazil, and Darwin had his first experience of a tropical forest. He was ecstatic and wrote a veritable paean to nature. "Delight . . . is a weak term to express the feelings of a naturalist who, for the first time, has been wandering by himself in a Brazilian forest. . . . To a person fond of natural history, such a day as this, brings with it a deeper pleasure than he ever can hope again to experience." Nevertheless, when much later he did again wander in a tropical forest for the last time before returning to England, he was just as deeply moved as on that first occasion.

The *Beagle* next sailed to Rio de Janeiro, where Darwin stayed ashore for three months, including traveling into the interior. From Rio, Darwin went on in the *Beagle* to Montevideo in what was then called the Banda Oriental ("eastern side" of the "Río," actually here an estuary, de la Plata). After a few weeks there, the *Beagle,* and with it Darwin, proceeded to Bahía Blanca, then a tiny settlement or military post to control or slaughter the Indians of the pampas. It is at the head of a considerable bay in the far southwest of what is now the province of Buenos Aires. At a place called Punta Alta along the shore of the bay about twelve miles from the settlement of Bahía Blanca, Darwin found numerous fossil bones and teeth.

The date of this discovery was in September 1832, on or shortly before the 22nd of that month. It was on that date that Darwin wrote in one of his rough notebooks, "On the coast about 12 feet high, and in the conglom.[-erate, a pebbly sedimentary deposit] teeth and thigh bone." Although this discovery and others similar to it were made by Darwin, it is fair to note that he was assisted by a young man named Sims Covington, who had been listed in the *Beagle's* crew as "Fiddler and Boy to the Poop Cabin." Darwin, however, trained him as an aid in collecting not only fossils but all sorts of natural history specimens. Darwin considered him a somewhat unlikable oddity, but kept him on as an assistant for several years after the end of the *Beagle's* voyage.

After that first visit to Bahía Blanca the *Beagle* continued cruising and Fitz-Roy continued charting. From mid-December 1832 to near the end of February 1833 they were on and around Tierra del Fuego, the big island (now divided between Argentina and Chile) cut off from mainland South America by the Strait of Magellan. Throughout March and into the first week of April 1833, they visited and charted the Falkland Islands. Although this takes us somewhat aside from our theme, in view of more recent events a brief excerpt from Darwin's account is worth quoting:

> After the possession of these miserable islands had been contested by France, Spain, and England, they were left uninhabited. The government of Buenos Ayres then sold them to a private individual, but likewise used them, as old Spain had done before, for a penal settlement. England claimed her right and seized them. The Englishman who was left in charge of the flag was consequently murdered. A British officer was next sent, unsupported by any power: and when we arrived, we found him in charge of a population, of which rather more than half were runaway rebels and murderers. [I add that these reprobates were replaced before long by highly respectable British subjects.]

In spite of his judgment of these islands as "desolate and wretched" Darwin did find a great deal to interest him there, but no fossils. Neither have any been found there since Darwin.

After some further voyaging, the *Beagle* landed Darwin at Maldonado, a small town on the coast of the Banda Oriental (Uruguay) near the mouth of the Río de la Plata and about sixty miles east of Montevideo. In the immediate vicinity is Punta del Este, now more widely known than Maldonado itself. In this region Darwin collected specimens of practically everything, including mammals but no fossil ones. The collections were laboriously packed for shipment to England, where Darwin's old friend and former professor Henslow was taking care of whatever Darwin could manage to send home.

The *Beagle* picked Darwin up at the end of July and sailed southerly to the mouth of the Río Negro, one of the five main rivers that rise on the slopes of the Andes and flow generally eastward across the width of Patagonia to the sea. A few miles up that river on its north bank was the tiny settlement Carmen de Patagones, at that time the most southern settlement of what we are inclined to call "civilized men" in the world. Since then a whole sequence of towns becoming cities has grown up along or near the Patagonian coast, and still farther is the now most southern city in the world, Ushuaia, Argentina, on the southern shore of Tierra del Fuego and the northern side of the Beagle Canal. Here a "canal" is a natural channel, not a man-made one. This one was of course charted by Fitz-Roy in the *Beagle*. The beautiful and imposing Darwin mountains are on Tierra del Fuego just north of the Beagle Canal.

At Carmen de Patagones, Darwin left the *Beagle* and proceeded by land northward across the Río Colorado. There General Rosas had established headquarters for the army, which was doing its best to exterminate the Indians of the pampas, a campaign eventually successful but still under way as Darwin and his escort went on to Bahía Blanca, which they reached on 17 August 1833.

From Bahía Blanca, Darwin for the second time turned to collecting fossil mammals at Punta Alta. In his published *Journal of Researches,* which as later revised and republished has become generally known as *The Voyage of the Beagle,* Darwin did not always follow strict chronological sequence, and his two different visits to Bahía Blanca and the collecting near there are mixed together in that text. Thus we do not know just when, or in some cases where, he collected what among the fossil mammals. For 1833 the rough notebooks indicate when and sometimes where he was collecting. On 21 August: "So tired of doing nothing started [from Bahía Blanca] to Punta Alta . . . Worked at cliffs & bones . . ." Some time between then and 23 August he must have visited Monte Hermoso, for his notebook here mentions fossil bones from what he miscorrectly called "Mt. Hermosa."

This point is of special interest because Monte Hermoso later yielded vitally important fossil mammals and has become the type locality for one of the land mammal ages for South America as a whole: Montehermosan in English; Montehermosense in Spanish. Monte Hermoso is not named in the *Journal of Researches,* but Darwin does there mention that at a time not given, "At the distance of about thirty miles, in another cliff of red earth, I found several fragments of bones." This was almost certainly Monte Hermoso, although if so the distance is somewhat underestimated.

After Darwin's return to England he edited and annotated a large work entitled *The Zoology of the Voyage of H.M.S. Beagle,* in which the identification and description of the fossil mammals collected by Darwin were assigned to Richard Owen, who was at that time the leading English authority on this subject. Darwin wrote a geological introduction to the section of this work by Owen. He was here more explicit about Monte Hermoso than he had been in the *Journal of Researches;* he clearly indicates that he both studied the strata at Monte Hermoso and collected fossil mammalian specimens from them. There is, however, one puzzling point here. Darwin says of the beds at Monte Hermoso that "even skeletons of animals, no larger than rats, have been perfectly preserved there." Yet no such skeletons were either described or figured by Owen in his part of the description of the zoology of the voyage of the *Beagle*.

In his book *Geological Observations on South America,* published in 1846, Darwin wrote at some length about Monte Hermoso (here thus correctly spelled out). In this he described "four distinct strata" and specified two beds in which, "especially in the lower one, bones of extinct mammifers, some embedded in their proper relative positions and other single, are very numerous." He noted that these included an extinct species of a genus, *Ctenomys,* of which a living species, called "tucutucu" in imitation of its call, survives in the same region, and another (there unnamed) extinct species of the living capybara "probably an inhabitant of fresh water." Capybaras, the largest living rodents, do live near fresh water and often enter it, but can hardly be said to inhabit it. They do not now live anywhere near Monte Hermoso, not extending south of far northern Argentina. Owen did describe and figure species classified as *Ctenomys,* considered distinct from the living species but not named. he also mentioned but did not figure or name bits of fossil capybaras considered to be of an extinct species. It is not surprising that later studies raise some doubts about these identifications.

In 1887, fifty-four years after Darwin, the great Argentinian paleontologist Florentino Ameghino sat at Monte Hermoso and mused about it. He noted that only its name is handsome—"Hermoso" means "handsome" or "pretty"—but that it is a place full of hitherto unknown life coming alive before our eyes as we dig up the fossils there.

In 1955, one hundred twenty-two years after Darwin and sixty-eight after Ameghino, I also sat and mused at Monte Hermoso. Indeed it is not "hermoso" and it is not a "monte" or hill in the common sense of the Spanish word. In this region "monte" usually means a thicket, brush, or open country not used for farming or grazing. The paleontological Monte Hermoso is a high, wave-cut scarp, and at its foot there is a wide platform (called a *restinga* in Patagonia), exposed at low tide and containing bits of fossil mammals. Here I had a feeling of awe, or even of piety, as it is a place made holy, in a proper sense, by the fact that those two truly great men had stopped and worked and mused here long before me.

The *Beagle* stopped in at Bahía Blanca while Darwin was there for this second time, but then went on with its charting while Darwin traveled north to Buenos Aires. He set out on 8 September 1833, accompanied by a reluctant gaucho whom he had hired. General Rosas had established posts where soldiers and horses were maintained as a means of travel and communication from the Río Colorado to the far distant capital. Darwin picked up a military escort on the way, came to a more grassy part of the great pampa with increasing numbers of cattle and some small settlements, and finally on 20 September entered the large city of Buenos Ayres (as Darwin always spelled it), then estimated to have some 60,000 inhabitants.

A week later, on 27 September, Darwin set out again overland with Sims Covington on a trip relevant to the theme of this book. They headed for Santa Fé on the Paraná river some 300 miles northwest of Buenos Aires. Not far south of Santa Fé at the stream called Río Tercero ("third river") or Saladillo ("somewhat salty") was the locality where, as mentioned earlier, Hugh Falconer had stated that he had found fossil bones. Grubbing around here Darwin found a fossil tooth which, to his great delight, later turned out to fit perfectly into a toothless skull to which Owen gave the name *Toxodon*. In fact the animal, one of the strangest and most typical in South America's extinct fauna, owes its name to this tooth. The tooth is strongly curved like a drawn bow, and in Greek *toxon* means "bow" and *odon* means "tooth."

A local man told Darwin that remains of giants occurred nearby in the banks of the Paraná. Going by canoe, Darwin found two groups of immense bones projecting from the perpendicular cliff. They were so completely decayed that Darwin could bring away only small fragments of a molar tooth, enough at least to show that the animal was some kind of mastodont. (Techniques for collecting fragile specimens are now highly developed, but were completely unknown then.) The local men said they had long known of such remains and had figured out why they were always so deep in the earth: like the local viscachas (or "bizcachas"), large as rodents go but hardly mastodonic, these big extinct animals must have lived in burrows!

The next day Darwin and Covington reached Santa Fé, where Darwin was ill for a time. Although at his then age energetic enough between bouts, he was often ill when ashore and always sick when at sea. When feeling better he crossed the Río Paraná to the small settlement of Santa Fé Bajada, just opposite Santa Fé itself but in a different province: Entre Ríos ("between rivers," the rivers being the Paraná and the Uruguay). Bajada, or Santa Fé Bajada, was then a city of some 6,000 inhabitants, a considerable size for so remote a place, and was the capital of Entre Ríos. (It is now called Paraná.) Darwin stayed there five days, enjoying the hospitality of an "old Catalonian Spaniard" and, once back on his feet, exploring the countryside for fossils.

North of Bajada, where d'Orbigny had found fossil remains, Darwin also found numerous bones and teeth. Among them were species later identified as *Mastodon andium, Toxodon platensis,* and *Equus curvidens*. Increased discovery and the vagaries of technical nomenclature have changed the exact designations, but there is no doubt that one was a mastodont, although now no longer put in the genus *Mastodon;* one was indeed of the genus still named *Toxodon;* and the third was certainly a horse close to or of the genus *Equus,* although other more or less contemporaneous genera of horses are now usually recognized in South America. This kind of association of a horse with animals long extinct had been indicated at Punta Alta and was made certain by the finds near Bajada. Even in the field Darwin saw that this association was sensational, and he discussed it in the *Journal of Researches.* It had been sufficiently established that there had been no horses in America, meaning all of North and South America together, when Europeans started to take domesticated horses there. Darwin wrote in the *Journal,* "Certainly it is a marvelous event in the history of animals, that a native kind should have disappeared to be succeeded in after ages by the countless herds introduced by the Spanish colonist!"

Although not often emphasized, this may be considered as the most important single result of Darwin's collections of fossil mammals during the voyage of the *Beagle.* In a different context Darwin stressed it in *The Origin of Species* at some length, as here quoted in part from the first (1859) edition. (No significant change about this had been made in the sixth edition, 1872, now more often reprinted and read.)

> When I found in La Plata [or more widely in what is now Argentina] the tooth of a horse embedded with the remains of Mastodon . . . Toxodon, and other extinct monsters . . . I was filled with astonishment; for seeing that the horse, since its introduction by the Spaniards into South America, has run wild over the whole country and has increased its numbers at an unparalleled rate, I asked myself what could so recently have exterminated the former horse under conditions of life so favourable. But how utterly groundless was my astonishment! Professor Owen soon perceived that the tooth, though so like that of the existing horse, belonged to an extinct species. Had this horse been still living, but in some degree rare, no naturalist would have felt the least surprise at its rarity. . . . If we ask ourselves why this or that species is rare, we answer that something is unfavourable in its conditions of life; but what that something is, we can hardly ever tell. On the supposition of the fossil horse still existing as a rare species, we might have felt certain . . . that under more favourable conditions it would in a very few years have stocked the whole continent. But we could not have told what the unfavourable conditions were which checked its increase. . . . If the conditions had gone on, however slowly, becoming less and less favourable, we assuredly should not have perceived the fact, yet the fossil horse would certainly have become rarer and rarer, and finally extinct;—its place being seized on by some more successful competitor.

Darwin connected this rather tenuously with his theory of natural selection, but of course he knew, and had stated clearly enough, that in this case, at least, "the more successful competitor" was obviously not the living species of horses introduced by the Spanish, as these never met with the horses already extinct by, as we now know, about ten thousand years. In fact Darwin did not know what "the unfavourable conditions" were in this case, and this is still one of the things that "we can hardly ever tell," although a good many unconvincing attempts to tell have been made. Just because we live at a later time, we can tell a good many things that Darwin could not, but this isn't one of them.

Now back to Darwin in 1833. Worried about catching up with the *Beagle* in Montevideo, Darwin took a boat down the Paraná to Buenos Aires, thinking that would be faster. It wasn't, and he found trouble in Buenos Aires, where General Rosas, tired of killing Indians, had decided to take over the whole government and had declared martial law. As an old acquaintance of Rosas from their meeting in Río Colorado, Darwin was able to get himself and Covington across the wide La Plata estuary to Montevideo, where he found the *Beagle* in port and in no great hurry to get away. Once more Darwin brought on board one of what Fitz-Roy called his "cargoes of apparent rubbish" although he later allowed that some of them proved to be interesting and valuable. Fitz-Roy can be allowed some complaint, for even though Darwin shipped his "rubbish" back to England whenever he could, there were long times when he could not, and he shared Fitz-Roy's cabin except for a short interruption when they argued about slavery, Fitz-Roy pro, Darwin con.

The delay in Montevideo gave Darwin one more chance for a land journey in the Banda Oriental, and he took advantage of it by leaving on 14 November 1833, this time for a loop to the northwest of Montevideo, to Mercedes near the mouth of the Río Negro, a tributary of the great Río Uruguay which by its junction with the Río Paraná forms the head of the estuary, so-called Río, de la Plata. Darwin had a grand time, lavishly entertained at estancias and enjoying the company and the activities of the gauchos. He also got some more fossil mammals. As he wrote,

> Having heard of some giant's bones at a neighboring farm-house on the Sarandis, a small stream entering the Río Negro, I rode there accompanied by my host; and purchased for the value of eighteen pence, the head of an animal equaling in size that of the hippopotamus. . . . When found the head was quite perfect; but the boys knocked the teeth out with stones, and then set up the head as a mark to throw at.

Darwin learned that the tooth he found near Santa Fé, many miles away and in a different country, exactly fitted into a socket in this head. Together, they became the basis for Owen's genus *Toxodon.* A few leagues from there (a league is roughly three miles) Darwin saw but could not collect parts of what was certainly a glyptodont, a big armored mammal the nature of which was then not really understood.

On 28 November Darwin was back in Montevideo and on 6 December 1833 sailed from there on the *Beagle,* never to return.

The east coast of southern South America, along with the Falkland Islands and much of Tierra del Fuego, had now been partially charted. The *Beagle* was now to go southward again, and then cross over to the Pacific for further charting in Chile. I should here mention that Patagonia, although an old name for much of what is now southern Argentina, has not been a separate country and is not now a legal or governmental unit. Darwin referred to the region of Bahía Blanca as in Patagonia, but the usage was dubious then and is not correct now. For the last century or so Patagonia has been generally considered as that part of Argentina from the Río Colorado in the north to the Strait of Magellan in the south. So delimited, it comprises four governmentally defined provinces, from north to south: Neuquén, Río Negro, Chubut, and Santa Cruz.

The *Beagle's* first stop on this cruise to the south was at Puerto Deseado ("Port Desire" to the English) on 23 December 1833. This is well down in Patagonia, within the province of Santa Cruz at 47°44′ south latitude. (Darwin noted it as 47° but it is nearer 48°.) Here Darwin first set foot in far southern Patagonia, and he found it very different from the region that had previously been his southernmost foray, Carmen de Patagones on the Río Negro. More southern Patagonia is limited in variety both of flora and fauna although the numerous guanacos, South American lowland camels, fascinated Darwin. An early attempt at a Spanish settlement had been made there but abandoned even before it was occupied. The *Beagle* party spent Christmas there. A guanaco shot by Darwin the previous day provided fresh meat, Fitz-Roy presented prizes, and everyone got drunk. Darwin found no fossils in this vicinity.

They left Puerto Deseado on 4 January 1834 and on 9 January reached the "fine spacious harbor of Port St. Julian"—Puerto San Julián—where they remained eight days. Here Darwin observed a basic geological formation containing gigantic oysters, up to nearly a foot in diameter, along with many other marine shells. Every geologist and paleontologist who has worked in Patagonia will recognize this as the Patagonia Formation, of vast extent and somewhat over twenty million years in age. Upward from the sea at San Julián there is a series of terraces each topped with heavy gravel which forms an even thicker bed at the top level. At one point "earthy matter filled up a hollow, or gully, worn quite through the gravel, and in this mass a group of large bones was embedded." These bones included most of the skeleton of one animal, but unfortunately they did not include the skull or any teeth.

From San Julián the *Beagle* revisited Tierra del Fuego for well over a month, including all of February 1834, and then also revisited the Falklands for about a month. Then the ship was taken back to the Patagonian coast for the principal purpose of exploring inland up the Santa Cruz river, the mouth of which was the next port south of San Julián. The *Beagle* had been here before, then under the command of Captain Stokes, and he had gone as far as thirty miles up the river. Fitz-Roy was now determined to explore it as far up as possible with three whale boats and a party of twenty-five men "sufficient to have defied a host of Indians," as Darwin recorded. There were Indians about, but they did not attack and only their tracks were seen. On 4 May 1834, they had reached the foot of the tremendous, snow-clad Andes, and Fitz-Roy decided to turn back. Most of the way up the river the men were towing the boats against a rapid current, and here rations were getting low. On 8 May, after just three weeks ashore, they were back on board the *Beagle*. They then took off for the Strait of Magellan, for the Pacific coast, and eventually after more adventures reached home at Falmouth on 2 October 1836.

In retrospect it is somewhat ironic that at San Julián, where Darwin collected one fossil mammal, and at Santa Cruz, where he collected none, he was in the general region, but not the closer vicinity, of what well before the century was out would be known to contain one of the world's largest deposits of fossil mammals, considerably older than any Darwin found. More will be said about this in a later chapter.

After the headless skeleton at San Julián Darwin never collected another fossil mammal. When he was back in England in 1836 his first regard, after reunions with family and relatives, was to sort out the collections, most of which had been sent back to John Stevens Henslow at Cambridge. The last shipment with fossil bones was probably sent from Valparaiso, the first considerable city after leaving Patagonia. Darwin was in and out of there for much of the latter part of 1834. Henslow had written to Darwin at Valpariso in July 1834 that the fossils (presumably most of the fossil bones) so far received had been sent to William Clift (1775–1849) at Surgeon's Hall in London for repair and preparation.

Much of the mass of Darwin's collections was duly written up by specialists in scattered publications, but the vertebrates, mammals, birds, "reptiles" (including amphibians), and fish appeared in the series of magnificent folios alluded to earlier on *The Zoology of the Voyage of H.M.S. Beagle . . . Edited and Superintended by Charles Darwin, Esq. M.A. F.R.S. SEC.G.S., Naturalist to the Expedition,* published by Smith, Elder & Co. in London, 1840–1842. The first section of this massive work was on fossil mammals, and after a geological introduction by Darwin it was by Richard Owen, Esq. F.R.S.

Richard Owen (Sir Richard after 1884), born in July 1804, was almost five years Darwin's senior. At the time relevant here he had been William Clift's assistant, became his son-in-law in 1835, and his successor in 1849. In 1856 he became superintendent of the natural history department of the British Museum in Bloomsbury, and it was largely through his efforts that this department became a separate museum, the British Museum (Natural History), and moved to a then new building on Cromwell Road, South Kensington, where it still stands. In the 1840s Owen was already beginning to be well known as an outstanding comparative anatomist and vertebrate paleontologist—a British successor to the world-famed Cuvier. He lived another half century, until 1892, and his long life thus began before and ended after Darwin's.

In his geological introduction Darwin specified the areas where he had found fossil mammals, already discussed in this chapter but not in just the chronological sequence of his discoveries: Bajada de Santa Fé near the Paraná; Banda Oriental (Uruguay, not more exactly specified); Punta Alta and Monte Hermoso near Bahía Blanca; and San Julián in Patagonia. To name each specimen or even species found by later collectors would become tedious and in fact impossible here, but for the Darwinian collection as published by Owen this is possible and worth while.

Toxodon platensis. Named by Owen and based on a cranium discovered in the bed of the rivulet Sarandis, this, well illustrated, is surely the nineteenpence target that Darwin bought from the boys. Its teeth are missing, as the boys had made certain, but Owen also provided a sketch of the single tooth that so elated Darwin when he found it far away at the Río Tercero in Argentina. In Darwin's collection from "Bahía Blanca" (doubtless Punta Alta) Owen also found, correctly identified, and figured a somewhat broken lower jaw of *Toxodon* that did contain most of the teeth. Owen's classification of this creature was extraordinary and will need some later mention. He called it: "A gigantic extinct mammiferous animal, referrible [*sic*] to the Order Pachydermata, but with affinities to the Rodentia, Edentata, and Herbivorous Cetacea."

Macrauchenia patachonica. This is the name devised by Owen for the headless and somewhat incomplete skeleton found at San Julián. Owen said of this, "an opinion as to the relation of the present species to a particular family of Ruminants, formed without a knowledge of the important organs of manducation, must be vague and doubtful, but the evidence about to be adduced, will be regarded, it is hoped, as more conclusive than could have been *a priori* expected." (For the benefit of those as ignorant as I was when I first read this, I note that "manducation" means "eating" but is now obsolete except with reference to taking the Eucharist, which would seem odd of *Macrauchenia*.) Owen's conclusion was that this "Mammiferous Animal" had "affinities to the Ruminantia, and especially to the Camelidae."

"Macrauchenia" means, in effect, "a big relative of the South American camels" as Greek *makros* is "big" and *Auchenia,* although derived from Greek *aukhen,* "neck," is a now invalidated name for the South American camels, llamas, guanacos, etc., which do have long necks.

Glossotherium. This was somewhat dubiously based on a broken piece of skull from the same Banda Oriental locality as the toothless but fairly complete *Toxodon* skull. Owen called it a new genus of Edentata, a group then supposed to include the aardvark of Africa as well as the armadillos and sloths of South America. Owen considered it nearly allied to the aardvark, which is now known to be grossly incorrect. Owen gave no specific name in the original description, but it was subsequently called *Glossotherium Darwinii* (or *darwinii* under the revised rules).

Mylodon Darwinii. This was described on the basis of an incomplete lower jaw, with all its teeth, found by Darwin at Punta Alta. Some confusion existed between the classification of this and of the hitherto exclusively North American *Megalonyx,* but Owen classified the South American animal as "a subgenus of megatherioid Edentata." It was clearly what is now called a ground sloth, as is *Megatherium,* but the genera are distinctly different.

Scelidotherium leptocephalum. This was based on a specimen, also from Punta Alta, and on the whole the best fossil mammal collected by Darwin. It comprises most of the skull and lower jaws as well as much of the postcranial skeleton, enough to have hazarded a reconstruction of the skeleton as a whole although this was not done by Darwin in the field, as has been shown in a fictionalized account, nor by Owen or Owen's artist on paper. There has been no serious dispute about the validity and application of this genus, but there has been confusion of the names *Glossotherium* and *Mylodon.* The two were closely related and similar, but somehow the names got mixed up. It was eventually decided, although not unanimously, that what paleontologists had been calling *Glossotherium* was really Owen's *Mylodon,* and vice versa. Such legalistic problems unfortunately are not uncommon in paleontology, although they do not involve what the animals really were. Owen thought that *Scelidotherium* was "allied to the *Megatherium* and *Orycteropus* [aardvark]." The former was not far off the mark, but the latter was. The resemblances that Owen saw between ground sloths and aardvarks were few and have proved not to be really significant.

"Megalonyx Jeffersonii." Owen so identified a ground sloth lower jaw found by Darwin in extremely poor preservation. He did not mention the locality. The identification was certainly wrong as to species and almost surely also as to genus, although *Megalonyx* has since been found in South America and had spread from there to North America at a late date, geologically speaking.

Megatherium. This largest of the sloths was already also the best known, as related in the preceding chapter. Darwin collected some more odds and ends of it, the newest feature of which was that one of his specimens showed a small fifth tooth in the upper jaw. It had previously been thought that the genus had only four upper teeth on each side.

Glyptodonts. These eventually massive armored relatives of armadillos had already been known from fragments, and a description of some of them had been published in Berlin by Eduard d'Alton in 1835. The name *Glyptodon* was given to them by Owen in 1839, the name meaning "carved tooth" from the strange pattern of the worn surfaces of the teeth. In the same year Peter Wilhelm Lund, who will be discussed in the next chapter, also described some specimens from Brazil. In his travels Darwin saw a number of partial or complete shells, but the knowledge and facility for collecting them entire were lacking, and he sent back only some loose scraps of the patterned armor, adding nothing significant to knowledge of them.

"Equus." The impression made on Darwin by discovery in two places of horse teeth associated with strange, extinct mammals has already been discussed here. Owen gave two figures of one of the teeth. In what could, but probably should not, be considered a somewhat condescending way Owen remarked that this was "not one of the least interesting of Mr. Darwin's palaeontological discoveries."

Ctenomys. The discovery of a possibly distinct fossil species of this still common genus of rodents ("tucutucus") is the only partial identification of odds and ends of smaller animals in Darwin's collections. There is no description in this work of Owen's of "the almost perfect carcasses of the several small rodents, the remains of which are so very numerous in so limited a space" at Monte Hermoso.

It was Darwin's impression that almost all the fossils he collected at such widely separate localities were of about the same age, geologically speaking. As far as identified, they all were extinct as species and mostly quite unlike any animals now living in South America although in a dubious way related to some of the latter. On the whole they were larger than any comparable more recent South American animals. They seemed to be geologically late in time, although the time in which they lived could not then be determined in years. Darwin did hesitantly suggest that some of the fossils from Monte Hermoso might be somewhat older. As much later determined, almost all his specimens described by Owen were Pleistocene in age and dated from between a million and ten thousand years ago, not a great span in geological history. Fossils from low in the beds at Monte Hermoso are in fact older, as Darwin suspected, perhaps as much as three million years old.

By 1844, when Darwin published his observations on geology in the parts of South America he visited, some further knowledge of fossil mammals there had already been made. For one thing a tooth and parts of the head of *Macrauchenia,* which had been lacking in the otherwise splendid specimen Darwin found at San Julián, had been sent to London from Buenos

Aires. From the tooth Darwin took this to be closely allied to the (much older) European *Palaeotherium,* which was wrong but no more so than Owen's thinking *Macrauchenia* was allied to the llamas. For another thing, more and more bones and even complete skeletons had been turning up in the pampas under conditions like those that produced the first *Megatherium,* discussed in the preceding chapter. Darwin now (1844) remarked that although "until lately, they [the Pampean fossil mammals] excited no attention amongst the inhabitants; I am firmly convinced that a deep trench could not be cut in any line across the Pampas, without intersecting the remains of some quadruped." (This prophecy proved to be correct; witness, for example the Botet Collection discussed in Interlude II.) Most important of all, a Captain Sullivan, R.N., whom I have not more closely identified, told Darwin that there was a formation rather high in the strata above Puerto Gallegos, ninety miles south of Santa Cruz, "highly remarkable, from abounding with mammiferous remains." This was almost certainly what is now known as the Santa Cruz Formation, an extremely good hunting ground for the Ameghinos, John Bell Hatcher, and others to be mentioned later in this book. I have noted that Darwin just missed them. It is presumably Captain Sullivan's finds that are mentioned as "not as yet . . . examined by Professor Owen" when Darwin wrote his book on South American geology. In fact after his description of the Darwin collection Owen did write a dozen or so other papers on South American fossil mammals, but all referred to Pampean fossils and genera already well known. It is amusing that one of these papers in 1841 was written to put his father-in-law (Clift) right on the point that it was *Glyptodon* and not *Megatherium* that had a "tesselated bony armour."

Although Darwin did not flatly state this at the time, it seems that by the end of the voyage the possibility that evolution has occurred had entered his mind and was being thought about, at least in an inchoate way. On 1 July 1837, less than a year after having returned to England, he "opened [his] first note-book for facts in relation to the *Origin of Species,* about which [he] had long reflected" (from his autobiography). As has already been noted here, Darwin had been interested in natural history from childhood, and it is conceivable that he would have followed this up and even have come to the knowledge of evolution without the voyage on the *Beagle.* Nevertheless, it was the voyage that made this eventually inevitable, that matured him, and in a sense professionalized him as an evolutionary biologist. That is true, even though Darwin's own specification of the things that were especially important in thus orienting him is open to question. In his autobiography he wrote:

> During the voyage of the *Beagle* I had been deeply impressed by discovering in the Pampean formation great fossil animals covered

> with armour like that on the existing armadillos; secondly by the manner in which closely allied animals replace one another in proceeding southwards over the Continent; and thirdly, by the South American character of most of the productions of the Galapagos archipelago, and more especially by the manner in which they differ slightly on each island of the group.

These observations surely quickened his attentive mind and were later to become integrated with his general view of evolution. They are therefore important in the history of human thought, but there is little reason to believe that it was just they that led Darwin to evolution and eventually to *The Origin of Species*. On the third point, for example, what came to be called "Darwin's finches" have been greatly and correctly publicized as ideal examples of evolutionary divergence in isolated populations from the same original ancestry. Yet when Darwin was in the Galápagos he did not spot this and did not even bother to note on what particular island each finch had been collected. Only later (especially in the 1845 *Voyage of the Beagle*, a revised version of the 1839 *Journal of Researches*) did he specifically consider the Galápagos birds as examples of divergent evolution. By that time he was far into evolution and was gathering as much evidence of all kinds as might bear on this one way or another. (Modern studies have confirmed that Darwin's second thought was correct.)

Our primary interest here is not with finches but with fossils, and here too there is a question as to how much contribution to later evolutionary Darwinism was really made by Darwin's collection, especially as it was treated by Owen. Taken only by themselves and with no more knowledge of fossils or of their stratigraphic succession, Darwin's collections of fossil mammals could not and did not lead directly to evolutionary conclusions. With the possible but improbable exception of the little tucutucu-like rodent, none of these fossils could reasonably be considered as ancestral, in an evolutionary sense, to any living species. Cuvier, who was and remained an antievolutionist, had demonstrated that *Megatherium* and *Megalonyx* had some definite anatomical resemblances to the living South American tree sloths, and from Darwin's collection Owen had added *Glossotherium, Mylodon,* and *Scelidotherium* to this slothlike list. Now, too, there was *Glyptodon,* anatomically somewhat like the living armadillos. But it seemed most improbable, and is now definitely known to be untrue, that any of these particular fossils were actual ancestors of their later relatives.

Darwin did make a point in favor of evolution from the apparent fact that these slothlike and armadillo-like creatures, living and extinct, occurred only in the Americas. (Darwin did not know that there were glyptodonts as well as armadillos in North America; there were, but that is beside the point here.) These animals could perfectly well have lived elsewhere, in

Africa for example, but apparently (and now we find this factual) they did not. That was a strong suggestion that they evolved in different regions. If they were specially created why were they not created everywhere suitable for them?

With *Toxodon* and *Macrauchenia* Owen's approach to classification led to absurdity. In *Toxodon* he saw a mixture of "affinities" with rodents, edentates—including sloths and armadillos among other groups—and "herbivorous Cetacea." (In present usage, and in Owen's at a later date, there are no such things as herbivorous cetaceans; here Owen was probably thinking of sea cows, which Cuvier had classified as cetaceans.) For Owen such "affinities," based on sometimes vague anatomical resemblances, were not evolutionary in what was destined to become the Darwinian sense of that word and process. To put things more briefly than can do full justice to Owen's complex discussions, he considered that each group of animals allied by "affinity" started from what he called an "archetype," a sort of transcendental, or perhaps supernal, overall model or pattern which was modified and diversified by a moulding force. I have been unable to gather clearly from Owen's works just what that force was thought to be and whence it was supposed to have come.

Owen's idea that *Macrauchenia* had some affinity with the living South American camels would make them for him results from the same archetype. For Darwin this was more evidence for what in *The Origin of Species* he called "the succession of the same types within the same areas." He gave Owen rather too much credit for illustrating this phenomenon on the basis of Darwin's collection of mammals, and it is just here that Darwin's South American collection is explicitly used in the treatment of paleontology in *The Origin of Species*. However, in this connection he stressed even more "the wonderful collection of fossil bones made by MM. Lund and Clausen in the caves of Brazil." That will be discussed in the next chapter.

Long subsequent study has shown clearly that *Toxodon* and *Macrauchenia* have no "affinities" to the groups with which Owen placed them. They belong to two very distinct major groups of animals, different orders in technical terminology. *Toxodon* was one of the last survivors of the previously enormously abundant and varied Order Notoungulata ("southern ungulates"), and *Macrauchenia* was the very last survivor of the also abundant but somewhat less varied Order *Litopterna* ("smooth heel," not a particularly enlightening name). Both of these large and long-enduring orders evolved in South America and are not known from anywhere else. Although Darwin could not know this, the history of these two orders is a splendid example of his "succession of types in the same area."

By 1859, and indeed well before the publication of *The Origin of Species*, a great deal more knowledge about paleontology had been amassed. The fact that Darwin paid close attention to this and devoted to it two of the fourteen (later fifteen) chapters of *The Origin of Species* may, after all, be credited to the fact that Darwin had himself early supplied some of the materials on which this progress was built.

One of the paleontological chapters (X in the first edition and 11 in later editions) is "On the Geological Succession of Organic Beings." Even in 1859 this was the most impressive evidence that evolution has occurred. Owen sometimes objected but did finally accept the succession as real, but he was among the then very few who did not accept Darwin's interpretation of it. There were gaps in the successions, and, although in many cases these have by now been filled in, the nature of geological deposition and of availability of fossils is such that an absolutely complete record can never be obtained. *The Origin of Species* treats this before it treats the record itself. original Chapter IX (later Chapter 10) is "On the Imperfection of the Geological Record." Here again the fact that Darwin had himself hunted for fossils supplied background.

After Owen had finished with the South American fossil mammals he wrote very extensively on other fossil mammals, and mostly on those from Australia. In his lifetime the known Australian fossil mammals were almost all of quite late age, geologically speaking, that is, within about the last two million years or so. He also continued studies of comparative anatomy, and in his position at the British Museum he had occasion to study many different kinds of fossil vertebrates from many countries, but not again from South America. He is honored in the history of science as a great descriptive anatomist, but he contributed little of lasting value to theoretical and philosophical aspects of biology.

Owen, who thus had a connection with South American paleontology through Darwin, became a famous man, and was knighted in 1884, when he was eighty years old. (He died eight years later.) Having had occasion to read much of his work and also to read some of his contemporaries' opinions of him, I may have become prejudiced, but I do feel that in spite of his talents he must have been rather unpleasant as a person. Sir Arthur Smith Woodward (1864–1944), who became Owen's successor as vertebrate paleontologist at the British Museum and whom I knew, had a low opinion of Owen. Darwin, who admired him in the first years after the *Beagle,* wrote in his autobiography that he "never was able to understand Owen's character. . . . After the publication of the *Origin of Species* he became my bitter enemy, not owing to any quarrel between us, but as far as I could judge out of jealousy at its success. . . . His power of hatred was certainly unsurpassed. When in former days I used to defend Owen, Falconer often said 'You will find him out some day,' and so it has proved." (As previously mentioned Hugh Falconer was a Scottish paleontologist who was an early discoverer and describer of the fossil mammals of India.) These remarks about Owen in Darwin's autobiography were among those suppressed by the family after Darwin's (and also Owen's) death.

Owen kept hinting or even saying that he could supply a far better theory of the sequence of organisms in the history of life than Darwin had done, but he never clearly said what his theory was. At times he seemed half convinced by Darwin, but if so he hid this under his hatred.

Notes

Anyone with a more general interest in Darwin should read at least the two most important of his books:

1845. *The Voyage of the Beagle*. London: Murray. (Also, later reprints of the first version, *Journal of Researches,* originally published in 1839 by Henry Colburn, London; and a facsimile reprint of the first version, published in 1952 by Hafner, New York and London.)

1859. *The Origin of Species*. London: Murray. (Also in facsimile, Harvard University Press, Cambridge. The 6th edition revised was published in 1878, and there are many reprints of it still on the market in various formats, some with new introductions.)

There are many biographies of Darwin, good, bad, and indifferent. A good one, authoritative, rather brief, and easy to read is:

Gavin de Beer. 1964. *Charles Darwin: Evolution by Natural Selection*. Garden City, NY: Doubleday.

The following is a readable and popularized rewritten account of the voyage of the *Beagle:*

Alan Moorehead. 1969. *Darwin and the Beagle*. New York and Evanston: Harper & Row. (This contains many fine full-color and black-and-white illustrations, not from Darwin's account. A grotesque "Reconstruction of the skeleton of a *Megatherium*" is said to be from *Journal of Researches,* which was Darwin's first edition of the *Voyage of the Beagle,* but it is not in my copy and is unlike any I have seen. The figures of bones of *Megatherium* and *Megalonyx* are reduced but good reproductions from Cuvier's *Ossemens Fossiles*.)

The following is the only full-length account of the life of Richard Owen that I know of. It is a typical Victorian family biography written by his grandson.

Rev. Richard Owen. 1894. *The Life of Richard Owen*. 2 volumes. London: Murray.

Lund; Winge

On 21 November 1954 Carlos de Paula Couto and I, accompanied by my wife and two local Brazilians, made our way into the Gruta da Lapinha, which in Portuguese means literally "grotto of the crèche." It is in fact a cave of considerable size with an arched entry in a vertical limestone cliff. Hence into increasing and then full darkness it continues deep into the earth as a series of chambers, many with interesting dripstone formations such as occur in almost all limestone caves. This particular cave has two special interests. One is that the limestone here is thinly bedded in layers of contrasting colors so that on some surfaces solution has produced intricate patterns resembling watered silk. The other special point, and the one that had brought us there, is that this is one of the caves, generally along the valley of the Rio das Velhas ("river of the old women") in which in 1835 and for years thereafter Peter Wilhelm Lund had collected an enormous number of highly varied mammalian fossils. Some scattered finds had been made before this, but it was Lund's long campaign that made the rich late pre-Recent (Pleistocene) fossil fauna of South America well and eventually widely known. We had come to this region as a sort of pilgrimage for me to see one of the *lapas,* as the bone-bearing caves are generally called there. Our companion and friend Carlos was the leading authority on Brazilian fossil mammals in general as well as on the life and work of Lund.

The Gruta da Lapinha is near—almost too near, as it is much frequented and littered—the town of Lagoa Santa ("holy pond"), a pleasant, somewhat resort-like and locally commercial settlement about twenty miles north of Belo Horizonte ("beautiful horizon," or somewhat figuratively "view"), a large but attractive city which is the capital of the state of Minas Geraes ("general, or varied, mines"). Belo Horizonte is some 220 miles north of Rio de Janeiro. The state of Minas Geraes is in southeastern Brazil, separated from the ocean by the intervening coastal states of São Paulo and Rio de

Janeiro. Although mining has somewhat fallen off, Minas Geraes owes its name, its relatively large population, and its variable prosperity to its remarkable mineral deposits, early and most profitably gold and diamonds, still producing many fine gems: aquamarines, topazes, and innumerable others. Most important of all, but not most profitable, is the fact that it also continues to produce fine fossils about a century and a half after Lund began his extraordinary work in this field.

Peter Wilhelm Lund was born on 14 June 1801 in Copenhagen. His father, a draper or cloth merchant, was of Jutland descent and his mother was from Ditmarsk, both areas Danish in population although the latter came to lie within the northern border of what is now West Germany. After his general schooling, Lund entered the Academy of Medicine in Copenhagen, only to decide a few years later not to practice medicine but to devote himself to natural history, especially botany but also zoology. In 1824 he wrote and defended two theses, one in medicine and one in zoology. The first, on medicine, involved animal experimentation, or, to put it more baldly, vivisection and the advances in physiology then recently made by these means. The second, although classified as zoology, had a somewhat related approach to study of the circulation of blood in "decapod crustaceans" (shrimps, crabs, lobsters, and the like). The two theses are said to have been written in a single year, 1824, which would seem to indicate both unusual mental and physical capacity. Each was awarded a gold medal; the first was translated into languages more widespread than Danish and was long considered the classic in its field.

There are differences both of opinion and of fact among the several notices of Lund's life, and this arises as regards his next move, which at first reoriented and later revolutionized his life. He had two brothers who died of tuberculosis, and one story has it that either for fear of contracting that disease, then usually fatal, or because he had in fact contracted it he decided to go to a more amenable climate and selected Brazil for that reason. Another version has it that he was tired of laboratories and offices and decided to pursue natural history where it really exists, in nature. Perhaps both motives affected him. In any case, he did sail to Brazil in October 1825, landing in Rio de Janeiro on 8 December after a rough voyage of six weeks.

He stayed three years in Brazil making collections and studing such diverse groups of animals as birds, ants, and snails, on all of which he wrote papers or memoirs. Besides excursions around Rio and elsewhere he spent fifteen months on a large ranch (*fazenda,* the Brazilian equivalent of *hacienda*). He sent back collections, especially to the royal museum of natural history in Copenhagen.

Lund returned to Europe in 1829 and for several years spent much time in travel, mainly to visit museums and universities to study their collections in natural history and to meet distinguished natural historians. At the University of Kiel he was made a doctor of philosophy with one of his Brazilian memoirs as his "inaugural thesis": a study of the genus *Eunope,* birds without a crop or craw. He visited Berlin, Vienna, Rome, Naples, Palermo, and Paris, among other places, and he met many explorers and naturalists, notable among them Cuvier, Humboldt, and Henri Milne-Edwards. The connection of Cuvier and Humboldt with South American paleomammalogy was noted in a preceding chapter. I mention the French naturalist Milne-Edwards (1800–1885) because it was he who particularly encouraged Lund to continue research in Brazil after Lund had apparently considered settling for a time in Paris to work over some of the results of his first visit. Lund may also have found the decision to leave Europe easier after the death of his mother, which in a sense broke his ties with his homeland. In any case he determined to go back to Brazil to study its zoology and more especially its botany over a larger area and in more detail. His reasons for returning may also have been those poetically put by the Brazilian geologist Henri Gorceix (in Portuguese, here translated):

> There are countries that are like marvelous books which, the reading of one page begun, restore our tranquility only after having been read to the end. How many others, less illustrious than Lund, have felt the same sentiment as regards Brazil!

So on 12 November 1832 Lund again embarked for Brazil, eager to see more of that country, but surely with no premonition that he would never return to Europe.

In Rio de Janeiro Lund joined a German botanist, Riedel, who had already been in Brazil for eleven years and knew the country, its language, and its customs well. They planned to take a long journey into the poorly known interior of the country, especially to study and collect its plants, in which Riedel was a professional and Lund was almost as well instructed as in zoology. Preparations for the trek took several months, but they finally left Rio on 18 October 1833. Their long botanical reconnaissance does not directly bear on our present preoccupation, and it suffices to say that one of its results was the joint publication in 1838 of a work the title of which, translated to English, is "Notes on the plants of the roadsides and wild herbs of Brazil."

On the return lap of the expedition, but still far from Rio, early in 1835, the two botanists (one of whom was about to cease being one) came to the settlement of Curvelo, which is between the Rio das Velhas, to the east, and the Rio São Francisco, to the west, but nearer the former. It is about 120 miles north and slightly west of Lagoa Santa, and thus not much farther from the main city of Belo Horizonte. Here again there are some differences among the various accounts of Lund's life, but they are not particularly important for our present purposes. It is certain that at about this time Lund met a fellow Dane named Peter Claussen who was known to the Brazilians as Pedro Cláudio or Pedro Dinamarqués, and it is probable that the meeting occurred by chance in Curvelo, not far from where Claussen had a *fazenda*. It is also probable that it was there and then that Lund heard of bones being found in the many caves, or *lapas,* of this general region. It is also certain that Lund and Claussen soon explored some of the caves together, although most of the collecting was thereafter done by Lund.

As soon as Lund learned of the abundant fossils in the lapas he saw that a great advance for science could result from their collection and study. Sometime in February 1835 Riedel and Lund parted company, Riedel going back to Rio with the botanical collections. Lund eventually settled in Lagoa Santa, at first only with the intention of collecting bones from the lapas fairly accessible from that town. It turned out that he remained there and thereabouts for the rest of his life.

In 1817–1820 Johannes Baptist von Spix, a German zoologist, and Karl Friedrich Philipp von Martius, a German botanist, had traveled widely in Brazil, and they wrote a tremendous publication on their travels there (*Reise in Brazilien*) published in 1831—posthumously for Spix, who had died in 1826. In it they described a cave known as Lapa Grande (meaning simply "big cave") near Formiga. This town, the name of which, oddly enough, means "ant" in Portuguese, is some distance southwest of Belo Horizonte and hence not in the region of the lapas explored and excavated by Lund, but with similar caves. Animal remains found in the Lapa Grande by Spix and Martius were ascribed, presumably by Spix, to tapirs, coatis, and *Megalonyx*. Tapirs and coatis still live in that region, and as far as I know it has not been determined that those found by Spix and Martius differ from the living species. Paula Couto has said that the identification of *Megalonyx,* a genus then known only from North America, was certainly incorrect but that the fossils so identified were probably either *Scelidotherium* or *Scelidodon,* extinct ground sloths not yet identified and named when the book by Spix and Martius was published. That, as far as I know, was the first report of fossils in Brazilian lapas before the collecting started by Claussen and carried on by Lund.

Even before Claussen and then Lund collected in the lapas, the presence of bones in them was well known locally by men who were exploiting the lapas not for the bones but for saltpeter, an impure nitrate useful as fertilizer and extracted from the yellow or reddish clay in which fossils were frequently embedded. It is probable that the buried animals contributed chemically to the formation of the nitrates. It is interesting that the more scanty remains of *Megalonyx,* as described by Jefferson and by Wistar, were first found also by diggers for saltpeter in the North American equivalent of a Brazilian lapa. It is further interesting but more coincidental that the ancestry of *Megalonyx* was South American.

Among the first lapas explored by Lund and among the richest in fossils was the Lapa Nova de Maquiné. This was the subject of Lund's first publication on the lapas and their fossils, a brief note published in 1836 in Danish by the Royal Danish Scientific Society. For the next ten years Lund continued to send such brief comments on discoveries in the lapas, written in Danish and generally appearing in Danish journals and periodicals, but occasionally reported also in French, German, or English. Translated from Danish, the title of his first extensive work was "Caves in the Limestone of the Interior of Brazil Some of Which Contain Fossil Bones." It was published in two parts in 1839 by the Danish Scientific Society. The first, longer part is devoted mainly to the Lapa de Maquiné, described in detail, with much appreciation of its beauty apart from the fascination of the fossils in it. Among its various rooms one with especially elegant stalactites was called "Castle of Fantasies."

The second part, or memoir, of this work is devoted to another cave known locally as the Lapa da Cêrca Grande ("cave of the big fence" or "enclosure"). A free translation of Lund's account of his approach to the cave will help understanding of this distant figure and his reactions to the world around him, more complicated and in a way even more appealing than his scientific passion for fossils.

> We were traveling to the south through a dense country forest, which became more and more thick. Suddenly the thicket opened and we saw before us a wonderful plain of rare and picturesque beauty. The borders of the forest were prolonged to right and left, forming an arc of a circle and encompassing the plain like a live hedge. In front of us rose a vertical wall of limestone which limited the plain in the south, going across from east to west.
>
> I imagined that I had before me the ruins of an ancient palace of giants, and my eyes were held in contemplation of a series of high arches cut into the left wing, as if in hopes of discovering there the vestiges of its mysterious inhabitants. The high top was covered with groves, gilded by the morning sun and peopled by innumerable flocks of golden-winged parrots (*Psittacus virescens*), whose strident cries, given vent to by our approach, denounced our coming to disturb them seriously in their remote refuge. A little *Cassia* with winged fruit attracted my attention. It was a new species for me and here it covered the whole plain, along with a *Melochia* with pale rose-colored flowers. . . .
>
> The admirable landscape that surrounded us had attracted the attention of primitive man. The nomadic natives—who I suppose were of the Caiapó tribe—had settled here, taking shelter in the caves of the imposing cliff. Inspired by the beauty of the landscape, they had

tried to picture the things in it, and the lower parts of the cliff are covered with drawings, which are, indeed, crude like the imagination that created them, but which do not fail to interest the philosopher who wants to know the products of the human spirit at the lowest level of its development. The Cliff of the Indians, near Ocambo, will always be a classic place for the traveling naturalist, in view of the extraordinary rarity of commemorative monuments of the savages of Brazil, such as this.

Lund then went on, here as in various other places, to find what he considered signs of the world-wide deluge that marked, in Cuvierian fashion, the last catastrophic change in the faunas and other features of the world. Then he went down into the darkness of the cave to seek among the stalactites and under the dripstone floors the remains of the antediluvian fauna, which so intrigued him as to be an obsession and yet, as the paragraphs quoted above bear witness, did not prevent his awareness of the beauty and interest of the present world.

Lund spent several days working in the Lapa da Cêrca Grande, which was fairly typical of the many in which he collected fossils over the years. Here, as usual, he found that the nitrate diggers had been destructive of fossils, but he was able to dig farther and to find several well preserved bones and teeth. In one place he found bones of a small deer, which he compared with the common living species that the Brazilians call "catingueiro," to which he gave the Latin name *Cervus simplicicornis* (now put in the genus *Mazama* as Paula Couto has pointed out). Elsewhere in the cave he found the jaw of some member of the dog family, in size between a fox and a wolf, and also a relative of the rodent called "paca" in Brazil, referred by Lund to Cuvier's genus *Coelogenus*. (By the vagaries of the code of technical nomenclature this genus is now called *Agouti,* although it is not the animal called "agouti" in vernacular English, "cutia" in the Brazilian vernacular.) Lund considered this fossil a species different from the living paca.

That is just one small sample of the toil that Lund happily carried on for some years.

Having described two lapas in some detail, with remarks on lapas in general and on the countryside, Lund began in 1837 a series of memoirs on the fossil faunas that he was accumulating. These were published in Danish by the Royal Danish Scientific Society between 1841 and 1845. Their general title or subtitle was "A Summary of the World of Animals in Brazil before the Last Revolution of the Globe." In reference to the history of animal life as known from fossils, Lund had become what might be called a disciple of Cuvier, and his running title for this series of memoirs was thoroughly Cuvierian. Having died in 1832, Cuvier did not live to know about Lund's accomplishments in South America and the interpretation of his finds in terms of Cuvierian catastrophism, with the Noachian deluge as the last of a series.

Those memoirs or monographs by Lund are of great interest, and they called attention to the richness and the peculiarity of the mammalian fauna of South America in the time of transition from ancient to recent. Nevertheless, they reflect the difficulty, indeed impossibility, of making an adequate and (even for the 1830s and 1840s) up to date study in Brazil of Lund's large collection. Brazil in that period lacked the laboratory, library, and other facilities for such studies even in Rio de Janeiro, let alone in the then remote back country where Lund chose to live and work. Such facilities for study of fossil mammals did then exist in Europe, notably in France and Germany, and they could have been developed in Denmark.

As late as 1848 Lund apparently was considering going to Denmark, or perhaps to France, and there making the serious, detailed, and for that period modern study of his fossils. Thereafter, however, these plans, or dreams, were abandoned. Already in 1843 he had written to one of his brothers in Denmark, "It would be madness on my part to think that I could stand our winter with my delicate heart and habituation to warm countries." One can believe that he really did dread the cold of a Danish winter, but it seems fairly obvious that this was also a handy excuse. As one biographer has put it, he was tired of his work, both physical and mental, and two psychological drives had become dominant for him: the resolves never to leave Brazil and to discontinue his scientific publications.

In 1845 Lund offered his collection to Denmark in a letter to King Christian VIII, whom Lund had known for years and to whom he now wrote, "This collection should be utilized for science as soon and as completely as possible, because of its interest and its scientific value." In fact, the making of the collection and ultimately its shipment to Denmark were largely, but apparently not completely, paid for by the king and by other subventions from Denmark. It is not clear to me how Lund made up the difference in expenses or how he lived for some thirty-five more years in comfort, always with at least one companion or servant, and noted locally as a benefactor to the poor of the village of Lagoa Santa. Presumably he inherited a comfortable amount, but about this I have found no statement.

The collection of fossils reached Denmark in 1849, destined for the Royal Museum of Natural History, but King Christian VIII had died in the previous year, the country was at war, and the museum lacked space. Lund's collection was crated up, piled in an obscure corner, and almost forgotten until 1859 when, thanks to an indignant newspaper article by a nephew of Lund, Trois Lund, the collection was unpacked and eventually found an honored place as the "Museum Lundii" in connection with the Zoological Museum of the University of Copenhagen.

Lund lived in happy retirement in his house in Lagoa Santa, reading, taking walks in the neighborhood, resting under a tree, often acting as host to European scientists who came by on their travels. Back in 1835 in the town of Curvelo he had employed a Norwegian named Brandt to be his assistant and right-hand man. Brandt also moved to Lagoa Santa and stayed there with Lund until Brandt's death in 1863. Thereafter a succession of young men, Danish, German, and Brazilian, helped to look after the aging and eventually nearly blind Lund until he died in 1880 at the age of seventy-nine. He was buried nearby in a place he had designated. Just fifty years later, the Brazilians having finally realized that he had been a great scientific pioneer, transferred his own bones to a new site and built a monument over the urn containing them. The Portuguese inscription on the monument states:

> Here lie the precious mortal remains, I say, remains of the illustrious and venerated sage, of regretted and imperishable memory, Pedro Guilherme Lund, born on the 14th of June of 1801 and deceased on the 5th of May of 1880.

Lund had lived long enough to know that his collection was finally in safe hands and that in his native Copenhagen there was a "Museum Lundii." He did not live long enough to know that his collection finally was "utilized for science . . . as completely as possible" as he had written to the king. Lund had even made some arrangements for the employment of a keeper or curator, namely his friend J. Reinhardt, to care for the collection. There was in fact a Johannes Theodor Reinhardt who between 1867 and 1888 published some studies in Danish on specimens from the Brazilian "bone caves," at least one of which was explicitly based on the Lund Museum. (There may be some confusion on this matter because elsewhere there is mention of studies on the Lund collection by a Joseph Reinhardt, but this is almost certainly an error for Johannes Reinhardt.)

However that may be, there is no question that the detailed and for its time the definitive study of the Lund collection was finally made not by a Reinhardt but by a Danish zoologist named Herluf Winge, who never visited Brazil. The son of an attorney, or the naval equivalent, in the Danish Ministry of the Navy, Winge was born in Copenhagen on 19 March 1857 and was thus only twenty-three years old and was still a student unknown to Lund when the latter died.

Even as a youth Herluf Winge made a collection of all the bones and teeth he could get hold of and unerringly aimed at a zoological profession in studying such things. He studied at the University of Copenhagen from 1874 to 1881, graduating as a "candidatus magisterii" in the terms of that place and day, later raised to the higher level of "magister scientiarum." In 1883 he became a member of the staff of the University of Copenhagen

Zoological Museum, first as an auxiliary, then as an assistant, and finally in 1892 as a "vice-inspector," which as far as I can judge would approximate an assistant or perhaps an associate curatorship in an American museum. He had no ambitions for titles and would not let his personal interests in zoology and in nature as a whole be interfered with by routine or administrative tasks.

Except for a tour of Italy and Switzerland after his graduation in 1881, Herluf Winge never traveled from Denmark, and rarely even left Copenhagen or its immediate vicinity for the rest of his life. His outings were usually confined to riding a bicycle around the suburbs, dressed formally as always in a long black frock coat and a big black hat with a wide brim. He never married, and he became more and more solitary after his brother Oluf's death in 1889. For some years he had been a regular attendant at scientific meetings with his brother, but after Oluf died he ceased attending. In 1910 he was elected to the prestigious Danish Society (or Academy) of Sciences, probably the highest recognition of his work, but he never attended any of its meetings either. Until his death, in 1923 at the age of sixty-six, he lived at the museum in working hours and otherwise in his small but handsome house with its adjacent area that he had converted into a haven for birds and small mammals. (In his will he left a large sum to the nation for the establishment of a preserve for the native fauna of the country.)

Despite his increasing reserve and eccentricity, visitors to his home were greeted politely, even cordially, if they did not unduly disturb his work. But intrusions were probably infrequent, because he was little known even to his colleagues in Denmark and had scarcely been heard of elsewhere. He wrote and published enormously, almost entirely on living birds and on both fossil and living mammals, but he wrote in Danish, a language practically unknown among scientists foreign to Denmark, and on subjects interesting to only a few or sometimes it seemed to no Danes. In this connection special note may be made of his relatively youthful monograph (he was then twenty-five years old) "On the change of teeth of mammals with special relationship to the shapes of the teeth." He then proposed a theory of the evolution of mammalian molars from those with three cusps in a triangle. This was similar to the tritubercular theory as suggested by the American E. D. Cope at about the same time or somewhat later. Neither Winge nor Cope had any knowledge of the other's work. Although the theory started students of the mammalian dentition in the right direction, the evolution of mammalian molars is now known to have differed from the relatively unsophisticated views of either Winge or Cope.

What here interests us most in Winge's study and writing is the large work labeled *E Museo Lundii* ("From the Lund Museum") which at long last came as near as could be to Lund's hope that his collection be studied "as completely as possible." This was issued in three large volumes, each in two or more parts, over the stretch from 1888 to 1915. It is more fully titled as: *A Collection of Monographs on the Animal and Human Bones Dug out by Professor Dr. Peter Vilhelm Lund from the Limestone Caverns of the Interior of Brazil and Conserved in the Paleontological Section of Lund in the Zoological Museum of the University of Copenhagen.*

The contents are best summarized by a list of the volumes and their parts or fascicles:

1888. Volume 1.
Preface by Chr. Lütken.
Part 1 by J. Reinhardt. "The bone-caves of Brazil and the animal remains found in them." A supplement lists the lapas worked by Lund.
Part 2 by Oluf Winge. "Birds of the bone-caves of Brazil."
Part 3 by Herluf Winge. "Fossil and Recent rodents (Rodentia) from Lagoa Santa, Minas Geraes, Brazil." [Winge regularly gave "Lagoa Santa" as the locality of the animals described, although none of the fossils and few if any of the living animals identified came from Lagoa Santa strictly speaking.]
Part 4 by Chr. Fr. Lütken. "Preliminary observations on the human remains of the Caves of Brazil and in the collections of Lund."
Part 5 by Soren Hansen. "The Lagoa Santa Race [of prehistoric humans]."

1893. Volume 2. (All by Herluf Winge.)
Part 1. "Fossil and Recent bats (Chiroptera) from Lagoa Santa, Minas Geraes, Brazil."
Part 2. "Fossil and Recent marsupials (Marsupialia) from Lagoa Santa, Minas Geraes, Brazil."

1895–1896. Part 3. "Fossil and living monkeys (Primates) from Lagoa Santa, Minas Geraes, Brazil."
Part 4. "Fossil and living carnivores (Carnivora) from Lagoa Santa, Minas Geraes, Brazil."

1906. Volume 3. (All by Herluf Winge.)
Part 1. "Fossil and living hoofed animals (Ungulata) from Lagoa Santa, Minas Geraes, Brazil."

1915. Part 2. "Fossil and living edentates (Edentata) from Lagoa Santa, Minas Geraes, Brazil."

Each of the sections except that by Reinhardt and those in Volume 3 had appended to it a summary in French. All of the sections by Herluf Winge include "an essay on the mutual affinities among the genera" of the group (in each case essentially a Linnaean order) being treated. These discussions were not confined to the fossil and living genera in what Winge treated broadly as the region of Lagoa Santa or even to those of Brazil or South America as a whole, but included those known, either fossil or living, anywhere and at any time. Later, in 1917 and 1919, Winge similarly treated two of the orders, Insectivora and Cetacea, not included in the Lund collections. Winge spent his last years, approximately 1918 to 1923, revising and assembling these essays into a treatise on the relationships and classification of all the known mammals.

This monumental work was published in Danish in three volumes under the title *Pattedyrslaegter* [Mammal Genera]. When Winge died in 1923 the first volume had been published, the second was in the press, and the third was complete in manuscript and could be published posthumously in 1924. Because of the linguistic difficulty, this culminating work of Winge's life was at first more admired, as a matter of faith perhaps, than it was read. However, A. S. Jensen, R. Spärck, and H. Volsöe made a translation into English which was published, also in three volumes, in 1941–1942 in Copenhagen and thereafter became widely known among all mammalogists. The English title is "The Interrelationships of the Mammalian Genera."

(I here allow myself a brief personal note. My full classification of the genera of recent and fossil mammals was published in 1945, but I had completed it several years earlier and had not been in a position to revise it after the publication of the English translation of Winge's last work. I had, however, slowly and painfully slogged through the Danish of some important passages in *E Museo Lundii* and *Pattedyrslaegter*.)

The isolated and rather eccentric nature of Winge's working habits colored his work on the Brazilian fossils and also his final masterpiece. Carlos de Paula Couto sums this up fairly and well in his introduction to the Portuguese version of Lund's work:

> Whoever contemplates all the vast scientific work of Winge cannot help paying him the tribute of profound admiration. Wherever one looks in the imposing assemblage of his publications there are found real mines of knowledge, all the richer the more concise it is—generally the author's style of writing in spite of its extraordinary clarity. This is especially notable in the monographs contained in *E Museo Lundii* and supplements, which beyond a doubt constitute the flower and the apex of the fecund litero-scientific production of this tireless naturalist. By this we do not mean to say that close criticism would not find any contradiction or fault. Where there is much light, there necessarily does not lack shadow. . . . Winge's phylogenetic deductions show above all an exaggerated Lamarckism. . . . But as the rediscovery of

> the fundamental Mendelian laws scarcely dates from 1900, it is evident that Winge grew up without the light cast by the new basic discoveries of modern genetics. . . . This lack naturally involves corrections of Winge's phylogenetic and systematic ideas without, after all, disparaging their unquestionable value.

To this I must add that while thoroughly agreeing in admiration for Winge, I cannot forbear noting that when Winge wrote *Pattedyrslaegter* genetics in the modern sense was established as to fundamentals and that Neo-Lamarckism had been rejected by most biologists. Winge's position on these points must be ascribed to his continuing and increasing isolation from the main currents of scientific thought. In a way this makes his accomplishment all the more remarkable.

The collaboration of Lund and Winge, two men who never met each other although their lifetimes overlapped, resulted in an important increase in knowledge of the Pleistocene and Recent mammalian faunas of South America. The field methods and records available for the Lund collection did not make a clear distinction between the cave specimens regarded as "antediluvian" in Lund's terms, and probably Pleistocene in geological terms, that is, about 10,000 years old or considerably more, and those geologically Recent, less than about 10,000 years old. Figures updated as of about 1950 or a year or two earlier were published in the work annotated by Paula Couto and cited in the following notes. These show 100 genera and 149 species recorded as "fossils" and 75 genera and 103 species recorded as living in the collections made by Lund and restudied by Winge. Most of the living genera and species are also recorded as "fossils," and one cannot be sure to what extent the supposedly fossil specimens (*fosseis* in Portuguese) were in fact recent (*atuais*) or were genera and species that had not evolved appreciably in the last thousand years, or a few thousands. In fact later, more modern studies to show that many or even most of the existing genera and species do date from that far back on the same continent in the same regions.

The greatest significance in these data is in the numbers of extinct genera and species in the Lund collections, which have been listed as 19 for genera and 42 for species. With few exceptions these quite probably are "antediluvian" in the usage of Cuvier and then of Lund, Pleistocene for modern geologists. Two of the exceptions are of special interest. Lund gave the name *Speothos*—from Greek *speos* ("cave") and *thos* ("wolf")—to a supposedly extinct genus of fossils of the dog family, but well before now it was found still to be living in South America, and not in caves. A kind of peccary identified as *Platygonus,* a genus well know from North America, is definitely long extinct there, but either it or something very like it quite lately has turned up alive in a remote and scarcely explored region in Paraguay. On the other hand *Equus,* the typical genus of the horse family, was identified in the cave faunas and was not counted as extinct. However, it was extinct long since in South America, as Darwin noted, and it should have been counted as antediluvian by Lund.

The point to be made here is not so much to discuss the nature of these faunas as to point out how much Lund and Winge, between them but not together, did add to knowledge of the latest South American faunas except the Recent. Much the most abundant living mammals of South America both in individuals and in genera and species are the field mice and the bats. These were practically unknown as fossils before Lund, but Winge turned up thirteen new species of field mice and Lund five. Among the bats, which might be expected in caves, there were twenty-seven species in the Lund collection, but these were almost all of living species. Of those considered extinct, Lund found only one species and Winge none.

Now in closing this chapter let us revert, or go forward, to 1954 when Carlos de Paula Couto, my wife Anne Roe, and I were spending some days in Belo Horizonte, not far from Lagoa Santa and some of the lapas. Carlos and I were working together on a monograph on the mastondonts primarily of Brazil but also of South America as a whole. We had gone to Belo Horizonte and thence to Lagoa Santa and a lapa near there because of an interest in Lund and also because we thought the university in Belo Horizonte might have some fossils from the lapas, perhaps including mastodonts. It was also on our way to a locality where a large collection of mastodont bones and teeth had been found. Most of that collection was now in Buenos Aires, where we had studied it, but some was still at a resort, Araxá, which was our destination. In the university at Belo Horizonte there was a collection of specimens from the lapas, including some excellent things but only rather puzzling bits of mastodonts. We found that this collection was purchased from the local British vice-consul, who had a hobby (illegal now, by the way) of cave- or lapa-crawling to collect fossils. When we met him, he let us study and photograph specimens he had not yet sold. These included a nearly unique skull of a young mastodont. Two of the same genus (*Haplomastodon*) had been found in Ecuador but were destroyed in a fire, and there is one in the Lund collection in Copenhagen, so the vice-consul's was one of only two then available for our studies. Carlos took four photographs of this specimen, which were duly published as Plate 18 of our monograph in the *Bulletin of the American Museum of Natural History* in 1957. According to the vice-consul this skull was found by him in the "Lapa da Caetano de Cêrca Grande," which I cannot more nearly place than as one of the many lapas in Minas Geraes. They are doubtless still yielding fossils, probably for sale, as were those of the vice-consul. Incidentally, he told us that he had refused promotion to more important consulates because, like Lund, he had formed a probably lifelong attachment for the country of the lapas.

Notes

Except for a few brief references I do not know any biographical data about either Lund or Winge in English. Even W. B. Scott, who was especially concerned with the fossil mammals of South America, did not mention either of them in the first edition of his classic book *A History of Land Mammals in the Western Hemisphere,* and he named each of them only once, in different connections, in the revised second edition of that book. In writing this chapter I have relied heavily on the book cited below, which besides Portuguese translations of almost all of Lund's publications contains likewise Portuguese accounts of his life, one by Carlos de Paula Couto and another by Henri Gorceix, who had based his in part, at least, on notes by J. Reinhardt, Danish like Lund and apparently his companion for a time in Brazil. There was also a summary of Lund's life and work published in 1943 in a Danish periodical *Dyr Natur og Museum* from which Paula Couto obtained some of his information. Paula Couto also provided notes, some of them extensive, on Lund's work in the same book as well as a summary biography of Winge.

Peter Wilhelm Lund, 1950. *Memórias sobre a Paleontologia Brasileira* Revistas e comentadas por Carlos de Paula Couto, Paleontólogo do Museu Nacional. Ministério da Educaçao e Saude, Instituto do Livro, Rio de Janeiro. This is a massive, paperbound book with 589 pages and 56 plates.

The English edition of Winge's last publication is:

Herluf Winge. 1941–1942. *The Interrelationships of the Mammalian Genera.* Translated by E. Deichmann and G. M. Allen. 3 volumes. Copenhagen.

INTERLUDE I

Muñiz

The name of Francisco Javier Muñiz is hardly known outside of Argentina, and even within that country it is not so widely known as that of Ameghino or, still less, of some military heroes and politicians. The name is, however, remembered and revered in Argentina by historians and by historically minded paleontologists. Among the latter Muñiz is known especially as the first native of what eventually became the Argentine Republic to have played a part in the collection there of fossil mammals. His role is exceptional in that respect, and to some extent it occupies a gap between Darwin and the Ameghinos. Thus it merits a place here in our history, but otherwise his contributions in this particular field are rather minor, and they are here inserted as a brief interlude.

This Muñiz was the son of Don Alberto Muñiz and Doña Bernardina Frutos, both of good families ("familias de viso"). He was born on 21 December 1795 and was christened Francisco Xavier Thomas de la Concepción Munĩz—"Francisco Xavier" in honor of the Jesuits, who had been expelled from South America, "Thomas" for Santo Tomás, as 21 December was the day of that saint, and "de la Concepción" from the pious belief, then not yet raised to a dogma, in the Immaculate Conception. This burden on the child was lightened by his later being called simply Francisco Javier Muñiz. His birth place was San Isidro, a village immediately northwest of Buenos Aires, now engulfed by that swollen city.

The first thing we know about his childhood is the astonishing fact that in 1807, approaching age twelve, he was a cadet in the Andalusian Regiment under the command of Colonel José Merlo. As a mere child, although in the army, young Francisco Javier was permitted to continue in school. Nevertheless, when the British attacked Buenos Aires for the second time in July 1807, he joined his regiment in the defense of the city. (The British were at war with Spain, and Buenos Aires was then a Spanish colony; in

1806 they took the city, but were forced out in about two months, and they attacked again in the following year.) With some of his fellow soldiers young Francisco fired on the advancing enemy from a flat roof. Then a group of them went down, "imprudently" as the report runs, into the street and opened fire less than half a block ("cuadro") from the advancing enemy. The British shot back, and the young fireball Francisco was hit by a bullet in the right thigh. Thus he showed the patriotism and the impetuosity that characterized his whole life. While his wound was healing he did have the satisfaction of knowing that the British lost the battle and failed to take the city.

Somehow Muñiz found time to study medicine and surgery, and in 1821 he was back with the army, assigned to the garrison in Patagones and thereafter to numerous other posts. In 1825 he was sent to the canton of Chascomús, on the pampa about seventy miles south-southwest of Buenos Aires. It was later recorded that sometime in that year he found and "disinterred" a number of fossil mammals, including "*daysipus jiganteus*" among others. That is an error or a misprint for *Dasypus giganteus,* the name given to a living South American armadillo by the French zoologist Geoffroy in 1803, now placed in a different genus, *Priodontes* Cuvier, 1825. It is unlikely that the fossil "*daysipus jiganteus*" was that species. It may have been a different armadillo or even a glyptodont.

In 1826 Muñiz was made "principal doctor and surgeon" by Bernardino Rivadavia, then president of the United Provinces of the Río de La Plata. In 1827, belligerent again, Muñiz headed the army medical corps at the battle of Ituzaingó (or Ituzangó), which is on the southern bank of the Paraná River, now the boundary between Paraguay and Argentina. Although apparently retaining connection with the government and the army, Muñiz was also made professor of "the theory and practice of childbirth, illnesses of infants and of the newborn, and legal medicine." That, like much else, was broken up during the tyranny of Don Juan Manuel de Rosas, who began his rise to power in 1829 and became an absolute and murderous dictator from 1835 to 1852. After the fall of Rosas, Muñiz was president of the Faculty of Medicine and was made a lieutenant colonel and principal surgeon of the army. He served at outposts and at the front in the battle of Cepeda on 23 October 1859, in which he was again wounded. However, raised to colonel, he also served in the bloody war with Paraguay, 1865–1870.

Somehow in the midst of what would seem an unduly occupied life, Muñiz in 1842 had announced that inspired by Darwin's fossil-collecting in the Banda Oriental and Argentina he was going to continue what he had started at Chascomús. He then made his base Luján and spent some years, more or less between wars, collecting in the general area where the first *Megatherium* had been found in the 1780s. The excavation from which it had been taken was still visible. Muñiz also found megatheres and many other fossil mammals, most important among them being a horse (an additional proof of horses being contemporaneous with megatheres as Darwin had found), and a great cat with saberlike canines which Muñiz somewhat conceitedly dubbed *Muñifelis*. Darwin, who was corresponding with Muñiz, suggested that the cat was probably *Machaerodus* (Darwin's error for *Machairodus,* a North American and Eurasian sabertooth "tiger"; the South American specimen must have been *Smilodon*).

In 1842 Muñiz had a rich collection of Pampean fossil mammals in eleven boxes which he presented to the government. The government then was General Rosas, who was more interested in "exterminating savage unitarians"—that is, catching and executing members of the political party opposed to his dictatorship. The boxes were shipped out of the country, and Muñiz long endeavored to find what had become of them. However, he himself followed up by offering collections of similar fossils to possible purchasers, including the Paris museum in the Jardin des Plantes, the Royal College of Surgeons in London, the Real Gabinete de Historia Natural in Madrid, and the Academy of Sciences in Stockholm. (The last-named received some of Muñiz's fossils as a gift and presented him with the Berzelius Medal.)

Muñiz did also present many fossils to the Museo Público de Buenos Aires, to the secretary of which, M. R. Trelles, he gave the following list in 1857:

> There figure among the organic fossils that I have placed in the Museum the magnificent head of *Toxodon platense* [italics not his] and other bones of this mammal. One of the hind legs of the Clyptodón [error for *Glyptodon*] and several of its caudal vertebrae. Another similar limb of the heavy Maghatherium [error for *Megatherium*]; the powerful arm and the terrible clawed hand of that giant of our ancient terranes, with some of its vertebrae and ribs; the sternum, clavicle, and istillar [word unknown to me], etc. etc. A mandible of the mastodont, whose molars preserve the enamel. A tusk or canine of the Mahamouth [error for "mammoth," but no mammoths have ever been found in South America], a species of elephant, which although imperfect in the end of its mandible nevertheless measures four feet six inches in length, retaining up to the end the natural enamel. A foot of Milodon [error for *Mylodon*], an animal of as extraordinary a form as those of those other species. Various bones of the fossil horse, etc.

Also in the *Escritos Científicos* ("Scientific Writings") of Muñiz there is an interesting letter from Charles Darwin, given in Spanish although undoubtedly written in English. (Darwin said he was not good at languages, and he probably did not write Spanish; I have not found this letter in the several volumes of his printed correspondence.) Muñiz wanted to sell some specimens, and Darwin advised him to send the fossils to some agent in England to be disposed of there. Among other things in this letter, dated 26 February 1847, was the following, which I am translating back from Spanish into English and which therefore cannot be just like Darwin's original:

> I cannot adequately express how much I admire your continued zeal, placed, as you are, without the means to pursue your scientific studies and without anyone sympathetic with you in the progress of natural history.
>
> I am confident that the taste for pursuing your tasks will be to you some reward for so much work.
>
> Some time ago you had the kindness to send to me by way of Mr. E. Lumb some *very curious* and for me *very valuable* information on the ñata cow.

(The "Mr. E. Lumb" mentioned in this letter was an Englishman resident in Buenos Aires. In 1833 while on the voyage of the *Beagle* Darwin had spent five days in the Lumbs' house, enjoying the comforts of a home away from home, with Mrs. Lumb pouring tea. It seems probable, but I have not found it stated, that Lumb translated into Spanish Darwin's letters to Muñiz and into English Muñiz's letters to Darwin.)

The ñata cattle, a word which Darwin spelled *niata* and which has gotten into English dictionaries in that form, were true-breeding mutants with pug noses, much like those of bulldogs. In 1868 Darwin published the information from Muñiz in *The Variation of Plants and Animals under Domestication*.

It is not surprising that Muñiz had information on the ñatas, an old breed dying out in Argentina. Apart from his profession of medicine and surgery, his enthusiasm for battles and the army, and his avocation of fossil hunting, Muñiz was interested in almost everything in the natural sciences and also in regional grammar and literature. In Darwin's letter here quoted Darwin also offered to send Muñiz his then recently published observations on the geology of South America, but he added, "I believe it would not be worth the trouble to send it to you without knowing whether you read English." Muñiz probably did not read English, as he knew Darwin's account of *The Voyage of the Beagle* only after it was published in Spanish.

Francisco Javier Muñiz died on 7 April 1871. In that year his most illustrious successor as an Argentinian paleontologist, Florentino Ameghino, was seventeen years old. Muñiz died as he had lived, dangerously and bravely. There was an outbreak of yellow fever in Buenos Aires, and Muñiz stayed there trying to do all he could as a physician until he, too, caught the fever and died of it.

This exiguous account of Muñiz may well end with a translation of the final paragraph by Domingo F. Sarmiento in a foreword to Muñiz's collected works. (When Muñiz died Sarmiento was president of Argentina; he lived until 1888.)

"If we lack the aptitude to constitute ourselves as the executors of his [Muñiz's] will, we can guarantee that we have more than enough conviction, and more good will, to go along the same trail that Muñiz followed as far as concerns the trail of a gaucho, of the description of the Argentine Pampa, and of the fine things that lie buried in it."

Notes

The material for this account of Muñiz is almost entirely from a small volume of *Escritos Científicos* of Muñiz published in Buenos Aires in 1916. Besides excerpts from some publications by Muñiz, this includes a long introduction by Sarmiento and shorter biographical appendices by Bartolomé Mitre and Florentino Ameghino.

Carlos Ameghino

At the beginning of this book I mentioned the town of Luján in Argentina and said that there are two reasons why all paleontologists should know its name. The first reason, as I went on to relate, involved the skeleton of *Megatherium* found near there, the first South American fossil to be classified and named properly and the first fossil skeleton in the world to be mounted as such and displayed. The second reason why paleontologists should feel reverence for Luján, although not as a rule piety in the usual sense, will appear in the present chapter and be further reinforced in the following chapter.

This part of the history starts in the city of Genoa in northwestern Italy when one Antonio Ameghino, who had lost his first wife and their daughter, married María Dina Armanino, also a Genoese. They went to the seaport Savona in 1853 and there took ship to Buenos Aires, where they arrived early in 1854 after a long and difficult passage. They moved at once to the nearby town of Luján, where they spent most of the rest of their lives together in a small house at 448 Las Heras street. There Doña María Dina Armanino de Ameghino gave birth to three sons: Florentino on 18 September 1854, a second son Juan at a date not specified in my sources, and a third son and last child, Carlos, on 18 June 1865.

In middle age Don Antonio became psychotic, for reasons not known, and he died at the age of fifty-eight in a Buenos Aires *nosocomio,* which in more usual Spanish could be called a *casa de locos* and in then usual English an "insane asylum." Of his three sons two were geniuses, each in his own way, and the other (Juan) was at least normally intelligent. Doña María, their mother, about whose origin and character I have been able to find practically nothing on record, lived to be seventy-six, an old age at that place and time. One can guess that she had unusual intelligence, which a woman could not develop and employ then and there but from the inheritance of which her sons profited.

It is the fact that both Florentino and Carlos Ameghino were born and spent much of their youth in Luján that makes this a place, one might say, of secular pilgrimage for paleontologists. There and in Mercedes, a nearby town southwest of Luján, Carlos acquired what little formal schooling he had, which amounted to becoming literate in Argentine Spanish. Florentino had somewhat more education of an elementary sort, in Buenos Aires, and he taught school for a time in Mercedes, where Carlos briefly joined him and had a final bit of formal education. The whole area around Luján and Mercedes, now much overgrown by the metropolis, was then mainly pampa and extremely rich in fossil bones. Florentino developed a veritable passion for them, and his kid brother, almost eleven years younger, also dug for fossils and developed a truly intellectual interest as well as a boyish mad enthusiasm for this activity. The family moved to Buenos Aires in 1867, where Don Antonio opened a small grocery (*almacén*), but this failed and Carlos and his parents soon went back to Luján. Carlos not much later briefly joined Florentino in Mercedes.

In 1878 Florentino went to Paris for an exhibition, taking his precious fossils with him, sold some of them, and used the proceeds to stay on in France until 1882, when he returned to Argentina bringing a French wife with him. His career will be followed more closely in the next chapter, which will center more on Florentino than on Carlos. In this period all three sons were most concerned with the support of their parents as well as of themselves and in Florentino's case of a wife. (They did not then or ever have children.) Florentino had invented a system of stenography which Carlos practiced and learned to perfection in the hope of employment as an official stenographer, which hope did not eventuate. Florentino suggested that instead Carlos might sell from door to door articles that Florentino would send him. Carlos, who was now eighteen years old, preferred digging for bones, and at about this time he found also some primitive implements in upper layers of the rich Pampean deposits. In the hands of Florentino, these were to lead far—too far, in fact, as will be seen later.

In 1886 F. P. Moreno, director of the then new Museo de La Plata, wanted more fossils to put in it, and he offered Florentino Ameghino the post of vice-director on condition that he exhibit there the already striking collection of Pampean fossils made by him and his younger brother. Florentino accepted and thereupon employed brother Carlos as *naturalista viajero,* that is as traveling naturalist or essentially as fossil collector for the museum. Carlos, delighted, set off forthwith for Patagonia and its fossils. This expedition in 1887 was to be the first of many.

It may here be added somewhat parenthetically that this was not strictly the first Ameghino expedition. In 1886 the government had sent a large group of naturalists under the leadership of Eduardo L. Holmberg, a distinguished zoologist, to make zoological, botanical, paleontological, and geological studies in the Chaco, the then poorly known extreme northern part of Argentina. Both Florentino and Carlos Ameghino were in the party, but it soon broke up and returned to Buenos Aires with little accomplished. As will be made clearer farther along, Florentino was a somewhat pugnacious and a touchy man always alert to sense an insult. Carlos, less prickly, was hardly capable of making enemies or nursing grudges, but became involved in some of them through loyalty to Florentino. It has been suggested that the fiasco of the 1886 expedition involved a quarrel between Holmberg and Florentino Ameghino, but this is unlikely. Florentino could not forget or forgive any personal falling out, but it appears that he did remain friendly with Holmberg for the rest of his life.

On the other hand there is no doubt that Moreno and Florentino Ameghino soon quarreled seriously and remained bitter enemies from then on. Early in 1889 Florentino was so incensed that he resigned as vice-director of the museum. When Carlos returned from his labors as *naturalista viajero* and found that his brother was no longer vice-director, he followed suit and resigned also. His expeditions continued, but now their expenses had to come out of the family income from two successive bookstores started by Florentino and eventually run by the other brother, Juan. The first of these *librerías* was in Buenos Aires and was known as "El Glyptodon" after the fantastically armored big Pampean animals of which Carlos and Florentino had already collected fine specimens. El Glyptodon also made and sold rubber stamps, paper, and some other odds and ends. (A *librería* is a shop where books are sold, not one where books are kept and can be read or borrowed: that is a *biblioteca*.) The Ameghino brothers later moved to the provincial city of La Plata and there also opened a *librería,* this one named Librería y Papelería Rivadavia. It was in the corner section of a one-story building also containing a house in which Florentino lived from 1889 until his death in 1911, Juan also until his death in 1933, and Carlos when not on expeditions. In his later years Carlos moved to Buenos Aires.

Let us return then to early in 1887 when the youthful Carlos Ameghino, not yet twenty-two years old, set out for Patagonia. Since Darwin, several other naturalists had visited Patagonia, notably the German Hermann Burmeister (1807–1892) and the Argentinian F. P. Moreno. These men were aware that there were fossils in Patagonia, but they did so little that it remained almost virgin territory for a ruggedly determined paleontologist, such as Carlos already was. Florentino already had a hypothesis, or perhaps as yet just a vague notion, that Argentina was the birthplace of many animals later dispersed more widely—and even of man as a species. This became an obsession, as will be discussed in the next chapter, and it was almost certainly one of his reasons for dispatching Carlos to Patagonia for evidence of world-important evolution having occurred there.

So Carlos sailed from the port (Riachuelo) of Buenos Aires on 25 January 1887 on the national transport ship *Villarino*. The ship stopped in for a week at Carmen de Patagones near the mouth of Río Negro. One of his shipmates, a Señor Jaca, otherwise unknown to fame, assured Carlos that he had encountered many fossils, probably megatheres and mastodonts, only three leagues from the town. Of course Carlos insisted on going there. He did so on the next day, only to find that the supposed fossils were not bones at all but mere concretions that had weathered out into what Carlos called "formas caprichosas" ("whimsical shapes"). Amateurs often mistake such concretions for fossils, but in this respect Carlos was already a professional.

On 14 February the ship reached the bay of Santa Cruz, another of the places visited by Darwin, as related earlier. Here Carlos disembarked and had to spend a week gathering pack and riding horses (a *caballada*), taking care of baggage, and generally preparing for a hard trip. On his first day, making a short start with some companions, he realized what he was getting into. He wrote in his diary:

> During this small trip I unfortunately have occasion to form an idea of the difficulties presented by carrying out a trip like the one I am going to undertake, making me foresee for the future great labors and fatigues, because, especially, of the scanty help I have, this being limited to a single man, since so far in spite of much perseverance it has not been possible for me to find here a skilled guide [*baqueano*] or even a servant [or apprentice, *peón*]. Carrying baggage on horses, I am now convinced, is much more difficult than I had previously imagined. I see that much practice is necessary in order to carry out the complicated system of ropes used to fasten and control that [the baggage] on the horse.

Carlos need not have worried. He would himself become the best *baqueano* and the best packer in all of Patagonia in those times.

After spending some time with settlers and traders near the bay of Santa Cruz and on islands close by, Carlos had his first night in the open, and although it was still summer, being February, which climatically corresponds with August in the Northern Hemisphere, he was so bitterly cold that he had no sleep. He then hurried back to a trader's and got a *quillango,* which is a fur blanket or robe made with the skins of young guanacos, the wild South American camels then still abundant in Patagonia. Thus more warmly equipped he went onward, generally to the west. He soon saw and recognized the marine beds, the Patagonia formation, also noted by Darwin, with its huge fossil oysters and other marine fossils but no mammals, and went onward with his two companions, Ataliba and Francisco. (He evidently had acquired another since he had complained of having been unable to employ more than one.)

His goal now was the Barrancas Blancas, steep slopes or cliffs of bare rocks, in this case white, which are often fossiliferous in Patagonia. (In other Spanish-speaking areas a *barranca* is a ravine, but the usage in Patagonia is different.) He reached them on 27 February, and he found in them the first fossil mammals since he had left the northern pampa at what already seemed a long time ago. Let us read a translation of his own description of this event, which turned out to be the beginning of a whole new phase of paleontology:

> After a light lunch, I make a visit to the closest barrancas. Walking along their bare flanks I have the luck to find some interesting remains of fossil animals, the first of a terrestrial [non-marine] sort that I find on this journey. In the hard strata of a little hill, separated from the main barranca, I collect numerous scutes [*tubérculos*] of the carapace of a *Hoplophorus* [one of the glyptodonts, armored relatives of armadillos], some parts of the head of a small edentate of the megathere family [i.e., a ground sloth], teeth of animals like *Toxodon,* fragments of the carapace of *Clamydotherium* and *Eutatus* [both are armadillos; *Chlamydotherium* is misspelled], teeth and jaws of various rodents, etc., etc. The satisfaction that I feel in being the possessor of such interesting objects is immense. Full of joy I walk to our camping place, forgetting for the time being the past fatigues and labors, daring to face others perhaps more severe in the future. For the night we spread our quillangos in the shelter that each has prepared beforehand, and I pass a pleasant night, abandoning myself to the meditation of my happy discoveries of this day.

Even in a personal retrospect of sixty years, a paleontologist feels the elation of young Carlos with one of the first of what will for him and his successors turn out to be thousands of such discoveries in Patagonia.

Carlos spent two more days at this locality and found further fossil mammals, which he recognized as similar to some found by Moreno in this general region, including some marsupials (not otherwise specified), *Interatherium,* and *Toxodontophaenus.* These names were apparently used in a lecture given by Moreno to the Sociedad Científica Argentina on 15 June 1882. The lecture had the odd title "Patagonia, remains of a continent now submerged"—odd because these were certainly land animals that lived in that region long after it had in fact been partially submerged. The text of this lecture is said to have been published by the society, but I have not been able to find a copy of it. It has been generally assumed that Moreno's lecture, published or not, did not technically define the genera and that they are not valid as of 1882. *Interatherium* was validated in the scientific sense by Florentino's publication in 1889 on the collection made by Carlos in that year. *Toxodontophaenus* (sometimes spelled *-phenus*) may have been a *Nesodon,* a name given by Owen in 1847 to a fossil probably from this same fauna. This was among the first episodes of a long and at times furious discussion over who named what and what name is correct for what group of fossils. I do not think we should attempt here to go into that maze farther than this example.

The party of three went on westward, sometimes along the river, sometimes on the easier going of the flatter surfaces of the mesetas at higher elevations. On 4 March they had the first distant view of snowy peaks, and the next day they could also see Lago Argentino, one of the largest and the southernmost of the beautiful lakes that lie along the eastern base of the Andes from southern Neuquén to here. Travel here was rough and hard on the horses. Furthermore there were problems in finding suitable camping places and in getting enough guanaco meat to supplement their supplies. On the night of 7 March it snowed heavily, and winter was obviously setting in here where they were at a considerable altitude and somewhat south of 50° in latitude. Carlos decided that this was as far west as they should go, and on the following day they turned east. On 10 March they came to large barrancas near a place called Yaten-huageno ("painted rocks") by the local Tehuelche Indians, a tribe then still abundant but now either extinct or no longer tribal.

The party spent several days near Yaten-huageno, and Carlos made a good collection of fossil mammals. On 12 March he dug out a complete skull of a small toxodontid that he had discovered the previous day. His account of this labor will bring back memories to any paleontologist:

> Provided with chisels and hammers I went my way even earlier than yesterday to the barrancas to dig out the magnificent skull discovered there. I managed to extract it in good enough shape, but it is impossible to imagine the immense work it cost me to get it out. The sand raised by a wind which blew strongly, producing a multitude of whirlwinds which blinded my eyes, and clouds of gnats anxious to bleed me bothered me horribly. Three hours of hard work and battling with those little but terrible enemies were necessary for its [the skull's] extraction.

When they were ready to leave, on 16 March, they had trouble finding their horses, which had been frightened during the night by pumas. (The puma, cougar, or mountain lion had the widest range of any single species in the New World, being found from well up in Canada all the way to southernmost Patagonia, although it has now been exterminated in parts of that range.) Carlos, "imprudently" as he put it himself (*una imprudencia mía*) set fire to the bushes near their camp, expecting the fire to go up the ravine and scare away the pumas, but instead it came down toward them, and they saved their lives only by "superhuman efforts." They then went forward to what Carlos called "the second white barrancas," which turned out to be richer in fossils than any place he had yet been. They stayed there until 21 March and then went on eastward with their already great load of fossils.

Back at the first "Barrancas Blancas" they spent about a week collecting still more fossils and picking up those left behind earlier. While there on 30 March Carlos made a geological, stratigraphic observation that was the most important of the whole expedition. In translation of his own record.

> Exploring the end of the barrancas I have made today an interesting discovery: I have observed almost at the level of the water or the river below the terrane of subaerial origin which contains fossils of land animals, [that] we have marine [rocks] very developed and with thick beds of oysters and other marine fossils.

This was what would eventually be called the Patagonia formation, and it was later mapped by Carlos for the whole length of Patagonia. The observation here clearly showed that the land fauna of the "Barrancas Blancas" was younger than the marine beds in this region, but that the land mammal beds here probably were not much later and in terms of geological time could have followed immediately after the sea withdrew from this region. It was already clear to Carlos and would be richly confirmed when Florentino studied, described, and named the fossils collected by Carlos on this expedition that this fauna of land mammals was different from and almost surely much older than the faunas already known from the pampas around where they were born and grew up.

The historical significance of this event at the "Barrancas Blancas" is that it was the first small step toward the tremendous achievement of determining most of the whole sequence of fossil faunas in Patagonia, in Argentina as a whole, and eventually in all of South America. For Patagonia especially, but also in part for other regions in Argentina, the working out of this sequence was done by Carlos Ameghino. He did this by the brilliance of his mind, the keenness of his eyes, and the persistence of his labors frequently under difficult conditions, without any previous knowledge of stratigraphy and other parts of the science of geology.

Let us now get Carlos back to Buenos Aires and then consider some of his subsequent work. He reached the coast and the Gobernación, or official government headquarters, near the mouth of the Santa Cruz river on 3 April 1887. He was occupied for some time in packing the fossil collection, writing, and waiting to be taken across to the north (or left) bank of the here broad river. On 14 April he started up the valley of the Santa Cruz again, but now on the opposite side of that river. He went through the region that had been previously explored by Francisco Moreno in company with Carlos Moyana. The weather was very bad, and Carlos found few fossils. He determined to have another look at San Julián, which was not too far away and where Darwin had found the skeleton of *Macrauchenia*. The weather continued bad, and near and at San Julián they found no water for themselves and for the horses. Carlos had no sooner reached San Julián than he turned around and headed back for the Gobernación, where he arrived on 15 May 1887. He was there for two months, expecting to return to Buenos Aires on the steamship *Magallanes,* but that ship was wrecked at Puerto Deseado. Three of the passengers made their way to the Gobernación of Santa Cruz, and couriers went overland to distant Punta Arenas, on the Strait of Magellan, to seek help for those stranded in Puerto Deseado.

Carlos finally did get back to Buenos Aires and to his family in La Plata. In April 1888 he made a short visit to Bahía Blanca and Monte Hermoso, again following in Darwin's tracks, but he found that those were not the only tracks there, as the following extract from a letter to Florentino shows:

> Dear Brother
>
> I have just come back from Monte Hermoso, where I never would have gone if I had had news of what goes on in that place.
>
> In fact since the visits of Pozzi and especially the last one of Moreno, who made here a real sweep, carrying off 19 crates of fossils, four people, as follows: the telegrapher, a lineman, the lighthouse keeper, and even his mother, spend all the livelong day on the beach, breaking any poor bone they see, which makes it very difficult to find anything worthwhile.

Nevertheless, indefatigable Carlos sent back to Florentino a collection that anyone else would have thought worthwhile: a well-preserved skull of a small ungulate, a poorly preserved one of a glyptodont, a "handsome lower jaw" of a carnivore, surely a marsupial although not so classified by Florentino, and about a hundred small specimens, mostly of rodents, some of them new.

In the meantime Florentino was working over the riches from Santa Cruz, which included more than 2,000 specimens. Florentino classified these in 83 genera with 144 species, almost all new. Even allowing for Florentino's usual inflation of numbers of Genera and species, that was a tremendous addition to knowledge.

In 1888–89 Carlos made his second major expedition to Patagonia, his last as *Naturalista Viajero* for the La Plata museum. Mainly in September through December, 1888, this took him over much of the territory of Chubut in central Patagonia, and he found some fossil mammals and also some dinosaurs, but this was not a very productive expedition.

Carlos continued going into the field, especially in Patagonia, every year until 1903, now at the family's expense. All of these expeditions were successful insofar as they provided materials for Florentino's studies and publications. Here, however, I will concentrate on some of the results of the expeditions from 1893 to 1900, which provided most knowledge, both faunal and stratigraphic, of ages greater than those of the rich Santa Cruz fossil fields.

As background here it must be noted that a Captain D. Antonio Romero had given Florentino a molar and a canine that had been found in Neuquén, the northwest territory of Patagonia. No more exact locality was given, and there were no data on the place of those teeth in the sequence of rocks with fossil mammals. Florentino recognized these as quite unlike anything else then known and gave them the name *Pyrotherium Romeroi* (by modern rules *romeroi*), as a new genus and species of perissodactyls, the order of mammals including living horses, rhinoceroses, and tapirs, along with many extinct groups. In the next year, 1889, Florentino put the genus in a family of its own, Pyrotheriidae, and in 1895 he separated it radically from other ungulates as an Order Pyrotheria. With increasing knowledge of the family, this isolation in a distinct order has almost always been maintained although there has been some dissent. In any case, the pyrotheres are certainly quite unlike anything known among living mammals or any fossil mammals from continents other than South America.

Early in 1894, after some further collecting in the region of Santa Cruz, Carlos made his way with difficulty northward to the Río Deseado, which is near the southern boundary of Chubut. Back at the port of Santa Cruz, he wrote to Florentino on 13 March 1894 at great length, giving him the exciting news of discovery of a fauna of fossil mammals older than any previously distinguished in Patagonia. As he went northward he found the Patagonia marine formation thinning out and generally overlying a floor largely consisting of porphyry and other ancient rocks without fossils. However, as he wrote:

> In the bottom of some deep depressions there are usually found resting on these ancient porphyries and sandstones limited lenticular deposits constituted for the most part by clays exclusively subaerial [i.e., nonmarine] in nature, given that they everywhere contain great quantities of remains of mammals. These deposits represent the hitherto mysterious epoch of the *Pyrotherium,* which turns out to be before the marine Patagonia formation. All this is based on direct and conclusive observations. These deposits seem to be the remains of an extensive terrestrial formation, perhaps destroyed for the most part by the invasion of the Patagonian sea and only appearing in view where the present large depressions chance to coincide with these small, ancient sedimentary basins. The deposits of this epoch which I have so far examined are insignificant, but doubtless the day that more extensive deposits are found there may be discovered a fauna as rich and numerous as that of Santa Cruz. From this point of view, north of the Río Deseado and [along] the Golfo de San Jorge now appears of much interest.

On maps the gulf of San Jorge looks like a tremendous bite taken out of the coast of Patagonia somewhat south of its middle. The city of Comodoro Rivadavia, a mere settlement in those days, is at the middle of the shore of this big semicircular gulf. What Carlos had in mind was to search the shore and eventually to search inland from there, which as we shall see he eventually did with great results.

Carlos's letter then goes on to give field identifications and some descriptions of the fossils found, including relatively abundant and unusually extensive parts of *Pyrotherium* such as the whole dentition, upper and lower. The fauna seemed broadly similar to that from the Santa Cruz beds, above the marine Patagonia formation, but was distinctly different in detail. Carlos remarked that rodents, numerous in the Santa Cruz fauna, were absent in his collection from the *Pyrotherium* beds. This was a local ecological or chance difference, as a considerable number of rodents have since been found in the upper-most part of the sequence then being lumped together as the *Pyrotherium* beds. Another distinction was that pyrotheres were abundant in the beds then named after this genus but were absent in the Santa Cruz fauna. After much further collecting that is still the case, and it seems that the whole group of Pyrotheria became extinct before the Santa Cruz beds were deposited.

In the summer of 1894–95, Carlos wrote to Florentino (8 May 1895) that among other things he had followed the shore of the gulf of San Jorge from Bahía Sanguinetti to Punta Casamayor, both on the southern coast of the gulf, in the Gobernación de Santa Cruz and well south of its boundary with Chubut. The bay and point designated by Carlos are insignificant topographic features not usually marked on atlas maps, but they appear on detailed local maps made long after Carlos's work there. As he indicated, the shore line between the Bahía Sanguinetti and the Punta Casamayor is about fifteen leagues in length. (In Argentina the league is equated by cartographers with five kilometers, which is quite close to the three miles usually taken as the length of an English league.) For that stretch Carlos found that what he was calling the beds of the *Pyrotherium* epoch were at the base of the sea cliffs, underlying the marine Patagonia formation. Carlos found these beds relatively poor in fossils in comparison with those along the Río Deseado.

The faunal and systematic studies by Florentino on the collections made by Carlos heightened interest in the Patagonian fossils. Florentino carefully omitted the exact localities from which the fossils came, but their nature and the general areas involved were becoming widely known. The Ameghinos remained alienated from the La Plata museum, but Carlos more or less kept track of its work in Patagonia and as far as possible Florentino did so in La Plata, where he was now living. Florentino's *bête noire* Moreno was still doing some collecting in Patagonia, and Carlos was beginning to think of this as poaching on his territory, although Moreno had collected some Santacrucian fossils before Carlos did. Santiago Roth started collecting in Patagonia for the La Plata museum in 1895 and continued there off and on until 1904, but his few studies thereafter were on the Pampean mammals, with one interesting study on a number of dentitions issued posthumously in 1927. Although he severely criticized Florentino Ameghino, they did sometimes exchange notes, but cagily and somewhat mendaciously. The classification of Roth's good collection of Patagonian fossil mammals and their places in geological time were not established until 1936, long after Roth's death. It is also highly noteworthy that the North Americans were coming. John Bell Hatcher collected for Princeton University in the marine Patagonian and the nonmarine Santacrucian beds in the three field seasons of 1896–1899, and for one season Barnum Brown was with him but collecting for the American Museum of Natural History (see Chapter 7).

After his not very fruitful coastal examination in 1894–95 Carlos wanted to extend his investigations inland in Chubut, but he was deterred in various ways for a few years. For example he wrote to Florentino from Santa Cruz on 12 January 1898:

> Recently I considered myself ready at the beginning of the year to undertake a trip to the gulf [of San Jorge], when misfortune willed that the only helper that I have should fall ill with the cursed influenza, and I had to postpone the trip up to now, but this date having arrived, I consider a trip to the gulf absolutely useless, given the great distance, and I believe that it is most convenient to be limited to make some exploration in more nearby places such as the Río Santa Cruz, etc.

On 19 April 1898 Florentino wrote from La Plata to Carlos in Santa Cruz. He mentioned that Moreno was more or less on the loose in Patagonia and wondered whether Carlos had had a run-in with him. Florentino went on to say:

> The fossils of the *Pyrotherium* epoch collected by the Museo [de La Plata] have certainly been found in the Río Chico del Chubut, near a deposit of dinosaurs. I have examined many of them and they have absolutely the same appearance and the same state of preservation, color, etc. as those that you found in [along the shore of the gulf of] San Jorge, being also the same species.

It is puzzling that Ameghino had examined specimens in the museum where he was decidedly *persona non grata* and still more so that he learned that they had been found "in" or along the Río Chico del Chubut.

Publication by Florentino on the *Pyrotherium* fauna had excited so much interest that a Dr. Valentin (otherwise unknown to me) had gone to Chubut in order to investigate that geological formation, but on the way, while climbing a barranca, he fell down a precipice and died instantly. Florentino added that other geologists and paleontologists, among them the Germans Gustav Steinmann and Ernst Koken, were going to Patagonia that year "with the same end in view." The end in view was certainly not that of falling off a barranca, and apparently they either did not come or came and found nothing, for neither of them published anything about such a trip or about Patagonian fossils. Somewhat later that year Carlos had information, which was correct, that Roth would be going to excavate for fossils at a locality near Lake Colhué-Huapí.

All this activity led Florentino on 30 November 1898 to write urgently to Carlos: "I recommend to you, very especially, the exploration of the gulf of San Jorge, that you undertake to determine the superposition and succession of the distinct beds from the coast to the lakes. This is just now most fundamental." It indeed proved to be so, and eventuated in what I consider the most important of Carlo's many field observations and collections.

As background for this, some geography of central Patagonia, especially in southern Chubut, is useful. Here, almost in the center of Patagonia, there are two large but relatively shallow lakes. They are in the meseta region about equidistant from the shore of the gulf of San Jorge to the east and the main chain of the Andes to the west. The western lake is officially called Musters, after an English explorer who was one of the first Europeans to travel near this region. The eastern lake is called by its Indian name, Colhué-Huapí, on almost all maps but is locally pronounced Coluapi, and so spelled in some of the Ameghinos' locality labels. Roth mixed up the names of these lakes, calling the western lake "Colhué" and the eastern "Musters," which for a time made rediscovery of his most important fossil locality difficult. (He said it was north of Musters, but it was really north of Colhué-Huapí.) Sarmiento, once a colony and now a town of moderate size, lies mainly between the southern ends of the two lakes, which are fed by the river variously designated as Seneguer, Senguerr, or Senguel. This arises from Lake Fontana, one of the Andean lakes, flows first east, then south, and finally curves around northward to the southern end of Lake Musters. When the water is relatively high it there branches over to Lake Colhué-Huapí, and when that basin overflows the water goes down the long course of the Río Chico (or more indicatively the Río Chico del Chubut). This river may be dry for much of its course, but when it flows it goes northeast and finally into the large Río Chubut above the settlements of Gaiman and Trelew.

South of Colhué-Huapí there is a large, high barranca, grandly scenic, known to paleontologists throughout the world as *the* Great Barranca. As one paleontologist (not to be coy about it, I) wrote, this "is the most imposing and important single known fossil mammal locality in South America, and one of the most important in the world. It must also be considered the greatest single discovery of Carlos Ameghino's extraordinary career."

In the summer of 1898–99 Carlos spent over three months in this region. After that field season he wrote to Florentino on 15 February 1899:

> It turns out that what we have been calling the *Pyrotherium* fauna according to new observations is, in reality, the succession of two different faunas, separated by an enormous interval of time, given that the strata that contain them are found to be discordantly stratified. The older of these two faunas, which might well be called of the *Notostilops* [correctly *Notostylops*], as that is the most characteristic and abundant genus, is indisputably Cretaceous since it is perfectly conformable with the Guaranitic terrane.

The point here is that there is indeed a fauna with *Notostylops* that is sharply distinct from the fauna with *Pyrotherium* and that is much older, by millions of years as we now know. It is not Cretaceous in age, and in fact the beds that Carlos was here calling "Guaranitic" are not Cretaceous either. He correctly did not include dinosaurs among their fossils. At this time Florentino was already convinced that the *Pyrotherium* fauna was Cretaceous and contemporaneous with dinosaurs, so he objected to the idea of an enormous interval between the *Notostylops* and the *Pyrotherium* faunas. This became deeply involved with Florentino's views, or *Idées fixes,* on the ages of these faunas and their relationships to others. That will be discussed in the next chapter.

Further collecting, especially in the Great Barranca, revealed that there is another distinct fauna between the *Notostylops* and the *Pyrotherium* beds of the Ameghinos. Carlos observed that in this sequence there are no evident disconformities, and this newly distinguished fauna was in the part of the sequence earlier thought barren of fossils. Carlos did not find many fossils in this part of the sequence, but enough to indicate a different fauna, dubbed the *Astraponotus* fauna by the Ameghino brothers. It turned out much later that most of Roth's fairly large collection from north of Colhué-Hupaí is of what had been called the *Astraponotus* age.

At the top of the Great Barranca there is one more level containing abundant fossils, more closely similar to the Santacrucian fauna but older, as they lie below and the Santacrucian lies above the marine Pantagonian. Four distinct faunas, all in one continuous exposure, thus lie one above another in this tremendous sequence. I known of no other place on earth where a single continuous transect cuts through such a sequence of fossil mammals.

In the valley of the Río Chico near its exit from Lake Colhué-Huapí there are exposures of late Cretaceous rocks containing dinosaurs; these lie beneath, and hence are older than, the rocks with the *Notostylops* fauna. Farther down the valley are richly fossiliferous beds with the *Notostylops* fauna, and a good part of Carlos's collection of that fauna came from here. He also found still farther down the valley a locality where extremely fossiliferous beds with a *Pyrotherium* fauna are disconformible with underlying beds that contain a rather sparse *Notostylops* fauna. The intervening *Astraponotus* beds are here absent as far as known. In 1911 a party from Amherst College led by Frederic B. Loomis spent some time at the locality and made an extensive collection of the *Pyrotherium* fauna, including the only skull so far known of the genus itself. (Loomis misinterpreted the geology of the area, as related below, in Chapter 6.)

Carlos continued working in Patagonia, and off and on in the lake and Río Chico regions, until 1903. In 1898 when he was just beginning his work in those regions he met there André Tournouër, who had been asked by Professor Albert Gaudry of the natural history museum in Paris to collect specimens from the interesting faunas that Carlos Ameghino was finding and Florentino Ameghino describing. Carlos reported to Florentino that he found Tournouër wandering at a loss because he had not yet been able to collect anything worthwhile and that he (Carlos) had told Tournouër to look in the barranca south of Lake Colhué-Huapí and along the Río Chico. Carlos was taken aback when Tournouër found new fossil mammals previously unknown to the Ameghinos. These were at first from near the top of the barranca and in the *Colpodon* fauna of the Ameghinos. Only in 1903 did Tournouër collect specimens of the *Notostylops* fauna, late in that year from low in the barranca and earlier from near Punta Casamayor on the coast of the gulf of San Jorge. These were not from Punta Casamayor itself, where Carlos had found only one not exactly identifiable specimen, but from a nearby ravine or small canyon ("cañadón"). It happened that both Carlos and Florentino were there at the time, and they called this locality Cañadón Tournouër in honor of their young follower—for by this time not only Carlos but more surprisingly Florentino had become friendly with Tournouër and helpful to him in his quest for fossils. What Professor Gaudry made of Tournouër's collections will be better considered later. (Parenthetically it may be remarked that in the 1930s nobody near what the Ameghinos called Cañadón Tournouër had ever heard of Tournouër. They called this feature the Cañadón de los Lobos—*lobos* means "wolves" in ordinary Spanish, but in Argentina it means "seals." By now the cañadón probably has still another name.)

It was purely by chance that Florentino Ameghino and André Tournouër met at Cabo Blanco, a small port at the southeastern end of the gulf of San Jorge. Florentino was arriving there on the transport *Guardia Nacional* with a preparator from the Buenos Aires National Museum, Emilio Gemignani, to meet Carlos, and Tournouër had just reached that point, having come from the Río Deseado. This was Florentino's first and only visit to Patagonia. As will be related in the next chapter, he had insisted that the *Notostylops* and *Pyrotherium* faunas were from below the Patagonian marine beds and were therefore Cretaceous in age—which has turned out to be a non sequitur. Others had claimed that these faunas were from above the Patagonian marine beds and were therefore late, even almost Recent in age—which was an even worse sequitur. Florentino wanted to see with his own eyes, and did see, that Carlos was right as regards the sequence, which did not prove that Florentino was right as to their geological ages.

That time with his brother in 1903 was the end of Carlos's last expedition in Patagonia, where he had spent most of his time for sixteen years. Admirers of Carlos Ameghino have perpetuated a legend that he made his explorations in Patagonia alone, on foot, with a back pack. It is true that he accomplished wonders in wild regions under very trying conditions, but both his letters and the mass of the specimens that he collected and sent back to La Plata make it obvious that he did not do this alone or on foot. He felt understaffed if he had only one helper with him, and he generally had two or more. He also traveled with a troop of horses (a *caballada*), up to forty or so of them, or sometimes of mules (a *muletada*).

After his retirement from Patagonia in 1903 Carlos did travel, but less extensively. He continued to make useful collections, some of them filling in the sequence after the Santacrucian, especially in the province of Buenos Aires but as far as Catamarca in the north. After Florentino died in 1911 Carlos published briefly on some of his later finds and finally, not quite so briefly, jointly with Lucas Kraglievich in 1921 and 1925.

While working at their own expense, the Ameghinos had sold small parts of their enormous collections, a few to the British Museum (Natural History) and to E. D. Cope (see next chapter), and more to the University of Munich, with which they had for a while an agreement that for financial support for their work they would give that university duplicate specimens in their collections. (The Munich collection was destroyed in World War II.) The great bulk of the Ameghino collection still in their hands was sold after Florentino's death to the Argentine government, which in the 1930s built a large new national museum of natural history far from the center of the city where it had originally been. Carlos Ameghino was for relatively short periods first head of the department of paleontology in the old museum in Calle Perú and later director of that museum. His most extensive biography says that he never entered the museum after 1930 and no longer had access to the Ameghino collection, but that is not true. He often visited the old museum in 1931, and thereafter had whatever access he wanted.

After essentially retiring from constant expeditions Carlos lived for a time with his brothers in La Plata. There Florentino controlled the home, to such a point that he prevented Carlos from marrying a young French girl—at least so it was later said by a not always reliable source. Some years thereafter Carlos did marry his first cousin, daughter of his mother's sister, Ascencia Merello Armenino, and they set up housekeeping in Buenos Aires.

In his late years Carlos suffered spells of "nervous depression" and was a semi-invalid, with occasional remissions during which he became more nearly his old good-natured and at times humorous self. Like both his brothers, he disliked doctors and rarely consulted one, to such an extent that the real nature of his illness is not clear. Apart from the vague depressions, he did develop arteriosclerosis. He died at 9 A.M. on 12 April 1936, in his seventieth year, having outlived his older brother by about a quarter of a century. He had no children, and this branch of the Ameghino family became extinct at his death, Florentino likewise having had no children and Juan never having married. These brothers did have two paternal uncles whose families continued the name Ameghino.

Carlos was short in stature, measuring only about 5 feet 4 inches, but in his youth he had been strong and rugged. He remained robust but did not become fat in his leisurely years. Although he lived more than thirty years after his sixteen years of Patagonian ventures, he retained certain Patagonian ways in speech and in habits. He was a great man, whose greatness was not recognized during his lifetime by anyone, not even himself.

Notes

The following is the nearest to a biography of Carlos Ameghino of any publication known to me. It must be said, however, that it is not well organized and that it is at least as much concerned with the ego of its author as with its subject, whom Rusconi speaks of largely as his "maestro" although their relationship was not in fact that of teacher and pupil.

Carlos Rusconi. 1965. Carlos Ameghino, Rasgos de su Vida y su Obra. [Carlos Ameghino, Characteristics of His Life and Work.] *Revista del Museo de Historia Natural de Mendoza,* 17:1–162.

The following work includes correspondence between Carlos and Florentino Ameghino, especially in volume 21:

Alfredo J. Torcelli, editor. 1913–1935. *Obras Completas y Correspondencia Científica de Florentino Ameghino.* [Complete Works and Scientific Correspondence of Florentino Ameghino.] 24 volumes. Buenos Aires: Taller de Impresiones Officiales.

For this chapter I have also had the advantage of being well acquainted with Carlos Ameghino in his sixties. While I was studying the collections he made, he spent many hours conversing with me and talking about his memories.

Florentino Ameghino

On a date that I did not record but that I believe was 6 August 1931 I was studying the Ameghino Collection in the old Museo Nacional de Historia Natural on Calle Perú in Buenos Aires. Days were ordinarily quiet in that old and rather decrepit building, but on this occasion I became aware of a growing stir in one of the halls. Investigating, I found that there was a formal assembly. There was of course, as is customary in formal assemblies, some speech-making, but what I remember most clearly was a group of young people singing something unfamiliar to me but sounding like an anthem. The words that caught my attention were repetitions of "Gloria, gloria a Ameghino!"

The sixth of August, 1931, was the twentieth anniversary of the death of Florentino Ameghino, and that is what makes me think that I do correctly recall the date of this gathering. The "Ameghino" of the anthem was of course for Don or, as he had become, Doctor Florentino Ameghino. There was no mention of Don Carlos, Florentino's younger brother, although Carlos was then living in Buenos Aires. He may even have been quietly attending the ceremony; oddly enough, I do not remember whether I saw him on that occasion. He was often in the museum, quietly visiting me in my work-room and drinking *maté* with me while he reminisced about Patagonia, from which I had recently come and which he had left a generation or so earlier.

It was always thus: *the* Ameghino was Florentino, with Carlos vaguely in the background for those who knew his name at all, and with almost no one aware that there had been a third brother, Juan. In fact, the great collection which was the basis of Florentino's glory had been made in most part by the relatively obscure Carlos, as recounted in the preceding chapter. That is why I have put Carlos first in this account. Juan's contribution was also significant but on quite a different level: it was largely his toil in the family bookstore that supported the scientific works of the other two brothers.

Florentino was much the most complicated and in personality the most striking character in the family. His fame grew and continued to spread in his later life and particularly after his death. Twenty years after that death not only were children singing his praises but also, as I soon learned, among nonscientists in Argentina any suggestion that Florentino had ever been mistaken was met with unbelief and resentment. And yet, as I must make clear in this chapter, Florentino was profoundly and consistently mistaken in more than one respect, admirable as he was in others. The Florentino worship has abated now among the growing number of his able Argentine successors in paleontology and even to considerable extent among Argentine laymen. The Argentine hero worshippers now have for their idols two Nobel Prize winners in science: Bernardo Houssay, physiology, 1947, and Luis Lelor, physics, 1970. Novel Prize awards started in 1901, and Florentino Ameghino was at his height thereafter in geology, paleontology, and anthropology, but there are no Nobel prizes in any of those sciences, and indeed the majority of twentieth-century scientists have not been and are not now eligible for Nobel prizes.

It is possible that Florentino's mother, Doña María Dina Armanino de Ameghino, was pregnant with him before her arrival in Argentina, and some of his detractors in later life even spread the rumor that he had been born before the family arrived, and therefore that he was not a native of Argentina. That is possible but improbable, and in any case could hardly detract from his fame even if true. The white population of Argentina by that time was, and now still is, largely of Italian origin, although all speak a variety of Spanish which they rather oddly call "Castellano." Florentino's birthday is usually taken as 18 September 1854, and his birthplace as Luján.

Florentino was both mentally alert and physically active from babyhood and onward until his terminal illness in his late fifties. In his earliest years there was no school in Luján, and it was said that he learned to read and write from his mother when he was six years old or even earlier. For a year an Englishman, not officially qualified as a schoolmaster, taught some of the young people in Luján. It may be one of the many tales of the precociousness of those who later became famous that after one year the amateur teacher told Florentino's parents that Florentino had already learned all that he, the Englishman, knew. Later a municipal school was opened in Luján, and Florentino studied there from 1862 to 1867; that is, approximately from his eighth to his thirteenth year, most of that time under one Carlos D'Aste, who was later honored for the education of his famous pupil. In 1867, aged thirteen, Florentino became "assistant" (*ayudante*) to the schoolmaster. In 1868 D'Aste arranged for Florentino to be enrolled in the Escuela Normal de Preceptores (essentially a training school for prospective teachers), but the school was soon closed for lack of pupils. Nevertheless, Florentino was officially given the title of "subpreceptor," a euphemism for "assistant teacher." That was the only nonhonorary academic title he ever held. (He was later given an honorary doctorate and held positions in a university and in two museums.)

Armed with that title, in 1869 the now fifteen-year-old Florentino was made "first assistant" in the elementary school in Mercedes, which, as noted in the last chapter, was another pampa village not far from Luján. It was probably in that same year that he began seeking and finding fossil mammals in the Pampean sediments along the banks of the Luján river. His first publications were probably letters to the editor of the local newspaper in Mercedes, *La Arpiración* ("The Ambition," an appropriate medium for aspiring young scientist). His first publication on fossil mammals probably appeared in the same newspaper in 1875, when he was about twenty-one years old. It was called "Notas sobre algunos fósiles nuevos de la formación pampeana" ("Notes on some new fossils from the Pampean formation"). No surviving copy of the original publication is known, but there is what purports to be its text in Volume II of the complete works (Obras Completas, cited here in the notes to Chapter 4). Most commentators consider this work as not technically published, a nicety of legal nomenclature that need not bother us here.

Something certainly was published around 1875, for contemporaries noted that at about that time Hermann Burmeister, director of what was then entitled the Museo Público de Buenos Aires, called Ameghino "an ignorant and pretentious youngster." Burmeister was then about sixty-eight years old. He lived to be eighty-five, and for the rest of his life he carried on a sort of running feud with Ameghino, both in and out of print. According to some scribbles found by his brothers in his notebooks, Ameghino either then or later retaliated, not in print, by calling Burmeister "the director of the Bible Museum" ("Museo Biblia"). This was the first of the bitter controversies that Florentino carried on for the rest of his life.

Burmeister was born and educated in Germany and was well along in a career as a paleontologist there before he moved to Argentina in the early 1860s. He was a Teuton of the old school and was a rigid biblical creationist who would have no truck with the Darwinian nonsense. Hence the significance of calling his museum the "Museo Biblia." Ameghino, on the contrary, early became a convinced Darwinian evolutionist. The two were still fighting this out in 1892, the year of Burmeister's death, when Ameghino published "Replies to the criticisms of Dr. Burmeister on some genera of fossil mammals of the Argentine Republic" (written in French but published in Argentina). Burmeister's most important activity in Argentine paleontology was that with contributions from Muñiz and others he built up the collections, especially of Pampean fossils, in the Buenos Aires museum. Florentino Ameghino benefited in this when he became in turn director of that institution, which he converted from a Museo Biblia to a museum Darwiniana, or eventually more specifically Ameghiniana.

In 1878 there was a grand exposition in Paris, and Florentino, now nearly twenty-four years old, went there, taking with him a number of his Pampean fossils. These were displayed, and many of them were sold, as Florentino apparently had no other source of money and was determined to stay on in France for some time. At that exposition he was given honorable mention and a bronze medal for his collection. (In 1882 there was a "continental exposition" in Buenos Aires, and his collections and works were given first prize and a gold medal. In 1889 there was another "universal exposition" in Paris which Florentino did not attend but where he was awarded a gold medal for his publications.)

During the 1878 exposition in Paris Ameghino met many distinguished scientists, with most of whom he corresponded after his return to Argentina. Among them was the American Edward Drinker Cope (1840–1897), who purchased part of the Ameghino collection. This went to the American Museum of Natural History when it purchased all of Cope's personal collection of fossils. Among the French scientists whom Ameghino came to know personally were the paleontologist Paul Gervais (1816–1879), the ethnologist with the elegant but florid name Jean Louis Armand De Quatrefages De Bréau (1810–1892), and the paleontologist Albert Gaudry (1827–1908). There were also the British mammalogist, Sir William Henry Flower (1831–1899), the Italian anthropologist Giuseppe Sergi (1841–1936), and many others.

Still in Europe after the exposition Florentino worked for a time at Chelles with Gervais (presumably Paul, although there was another member of the family, Henri, with whom Florentino wrote a book in French on the fossil mammals of Argentina, published in 1880). Chelles was, and still is, a famous locality with evidence of prehistoric (Pleistocene) man, type of the Chellean age of archaeologists. Florentino then traveled widely, visiting museums in France, Belgium, England, and Italy.

Sometime while in Paris Florentino married a French woman, Léontine, later Argentinized as Leontina, Poirier. When Florentino returned to Argentina she went with him and lived there for the rest of her life. She died in 1908, three years before her husband. Her date of birth does not seem to be on record. In 1895 Florentino named a genus, *Leontinia,* and a family, Leontiniidae, in her honor. (A cataloguer who should have known better recorded that *Leontinia* was named for "Leontina——, a friend of Dr. Florentino Ameghino.") They had no children. After Florentino's death one of his eulogizers wrote that famous people rarely have children because if they had any this would interfere with their work. Someone should have told him that Charles Darwin, who was emulated by Florentino Ameghino, had ten children.

Before he went to Europe Florentino had learned to read and write (probably also to speak) French and Italian. Most of his publications were in Spanish, but some of the most important were in French, which in his time and milieu was the most nearly international language. It has been suggested that Leontina may have helped in writing his French manuscripts, but this is unlikely. Her native French was apparently more colloquial than literary. Florentino and Leontina rarely had occasion to write to each other, but a note from Leontina to her husband has survived. It is in Spanish but its salutation is in peculiar French: "Cher petits maris cheris."

Florentino wrote in Italian to his Italian correspondents but did not publish in that language. His North American and English correspondents generally wrote to him in English, but he usually replied in French. Cope at least once wrote to him in French, but that was exceptional. His German correspondents usually wrote to him in German, and he usually replied in French. An exception was the outstanding German paleontologist Karl von Zittel, who wrote to Florentino Ameghino in French. The German paleontologist-zoologist H. von Ihering, who spent much of his life in Brazil, was a frequent correspondent of Florentino Ameghino's and wrote to him in Portuguese. Florentino replied in Spanish. It is clear that besides French and Spanish Florentino could read English, German, Portuguese, and Italian, but he did not write or publish in those languages.

When Florentino returned to Argentina in 1882 he apparently expected to carry on as principal ("director") of the municipal school in Mercedes, but he found that he no longer held that position. He had been off on his own in Europe since 1878. One would think that the loss of his job would not be a surprise, but one of his elegists (Victor Mercante) thought otherwise: "Fortunately there was in Ameghino an excess of fortitude, moral strength, indeed not to be uncrushed but not to break out in affronts and to speak out against the unjust decision that left destitute a master because he had from the other world projected an aspect of glory, the first for an Argentine savant, on his country."

"The other world" is apparently meant for Europe, where Ameghino had indeed made himself known to numerous European savants, and at least one American. He had also, as joint author with Henri Gervais, written the book mentioned earlier on *The Fossil Mammals of South America* (Buenos Aires and Paris, 1880). Further, he had written an immense two-volume work in Spanish on *The Antiquity of Man in La Plata,* published in 1881.

Florentino, driven not only by ambition but by the ideas pullulating in his mind and demanding expression in written words, was virtually destitute. It was then that he opened the bookshop named for the armored, club-tailed Glyptodon, second cousin to armadillos. There he brooded, speculated, and wrote while earning a living as shopkeeper. He thought in particular about the ungulate fossils known as *Toxodon* and *Typotherium,* which found no place in the classifications then current, and he concluded that the fault was not that these animals could not be classified but that the classifications were wrong even in principle and had to be replaced. As to this he later wrote (in Spanish):

> Thus was born *Filogenia,* in which should not be seen a literary work inasmuch as, seeing myself obliged to obtain my daily bread by taking care of my bookshop business, I write every line between the sale of four reals-worth of pens and one peso-worth of paper, a situation little favorable for giving my ideas lofty literary forms.

That book, *Filogenia* (phylogeny), published in 1884, had as its subtitle (also in Spanish) "Principles of evolutionary [*transformista*] classification, based on natural laws and mathematical proportions." It must be admitted that this ingenious and thoughtful book has had almost no influence on subsequent taxonomy (the principles of classification of organisms) except in the work of Florentino Ameghino himself. It did show that the orders of mammals as then named, largely but not entirely still following Cuvier, were not all well defined and in particular that a number of the fossil South American mammals did not fit into any of the earlier adopted orders. Ameghino did not provide here a new overall classification of mammals, but he did briefly discuss the two genera about which he had been brooding, *Toxodon* and *Typotherium*. He placed them both, along with some other genera rather clearly related to one or the otther of them, in a group that he informally called "los pentadáctilos." This means "five-fingered," but obviously was not meant to be taken literally, as Ameghino excluded a great many genera with five fingers, *Homo* for instance. He wrote that through *Toxodon* the "pentadáctilos" were mixed up or confused with the pachyderms and through *Typotherium* with the rodents. "Pachyderm," a term no longer in use, then included the larger living hoofed mammals. Incidentally, by a quirk of technical nomenclature the correct name of what was long called *Typotherium* is *Mesotherium*. Later, especially in his great work of 1906, Ameghino evaded the issue by putting *Toxodon* and its allies in a distinct order Toxodontia proposed by Richard Owen in 1858 (now usually written Toxodonta) and *Typotherium* (= *Mesotherium*) in a distinct order Typotheria, proposed by Ameghino's correspondent Zittel in 1892.

It is one of the ironies of history that Santiago Roth, the generally unsuccessful rival of the Ameghinos, essentially solved the problem that so bothered Florentino. In a work called "Los ungulados sudamericanos" ("South American ungulates") he showed that most, although not all, of the then known extinct hoofed mammals of South America, including toxodonts and typotheres, could be put in the same major group, now classed as an order, named by Roth Notoungulata, a latinization of "southern ungulates." This is now almost universally accepted.

Now let us return to Florentino Ameghino sitting at his desk in his shop called El Glyptodon. The *Filogenia* was a learned book in a developing country that was more concerned with politics and warfare, external and internal, than with learning. In 1880 General Julio Roca, in command of the provincial forces centered on the city and province of Córdoba, attacked and took Buenos Aires, which was thereafter federalized although still in fact under control of the Córdoba League. The then minor city of La Plata was made capital of the province of Buenos Aires. Córdoba long continued to be the center of political power and also of learning: the National Academy of Sciences was established there. The relevance of this here is that the *Filogenia,* although it was little read, was noticed as an unusual scientific production. A result was the appointment in 1884 of Florentino Ameghino as professor of natural history in the University of Córdoba. He was also made an honorary doctor there, and for several years published papers in the bulletin of the (Cordoban) National Academy of Sciences.

In 1886, however, as already related in Chapter 4, he moved to La Plata, as vice-director of the museum there and director of the paleontological section. The falling out with the museum's director, F. P. Moreno, which occasioned his resignation only a year later was not the first or the last of the bitter disagreements—or feuds—between Florentino and various associates. Like most of the others it lasted while both of the antagonists lived. In this connection it may be mentioned that the British naturalist-paleontologist Richard Lydekker (1849–1915) visited Argentina in the 1890s and wrote a note to Ameghino, not quite apologizing, saying that as he was the guest of Moreno he had been unable to visit Ameghino. Lydekker collected some fossils in that country and published on Argentine fossil vertebrates. Those publications were sharply attacked by Florentino Ameghino in 1894 and 1895.

It should, however, be added that Florentino was quite amicable to some whom he could have resented as rivals or as critics: his relations with André Tournouër or with W. B. Scott, noted elsewhere in the book, bear this out. While on the subject I may also add that Carlos Ameghino, usually more even-tempered than his more famous brother, in 1915 did react strongly and adversely to a book by Frederic B. Loomis on *The Deseado Formation of Patagonia,* published in 1914, three years after Florentino's death. In the light of personal experience, I agree thoroughly with Carlos on this subject, as my further account in Chapter 6 will make clear.

Having left the La Plata museum in 1887, Florentino and his brothers stayed in La Plata and opened another bookshop there, as also noted here in Chapter 4. In 1902 Florentino was made director of the national museum in Buenos Aires, successor to Burmeister and then Carlos Berg. There was some objection to this appointment. Although he was known worldwide within his profession, that profession was hardly known in his country. He kept much to himself there, with a low profile, quite remote from the political and martial arts that were then the usual paths to worldly success in Argentina as elsewhere. He was the obvious choice for the directorship, and yet that position was not entirely congenial to one of his temperament. He worried about the museum, and then buried those worries and himself in reading and writing. Except for his one short expedition to Patagonia in 1903 with Carlos, he had little physical but much mental activity in the fading years. He longed to move the museum's collections into a larger, more modern building, but could achieve no progress at all toward that end. The museum he dreamed of was finally under way twenty years after his death, and then it was not named for him, as many had suggested. Instead it was named for Bernardino Rivadavia (1780–1845), who figured in the war of liberation from Spain, became governor of Buenos Aires and president of the Argentine Federation, resigned in 1827, and spent most of the rest of his life in exile. I find no record that he had any interest in natural history or in museums.

In his last years Florentino had severe and at that time incurable diabetes mellitus, which dragged on with increasing debility and severe pain until death relieved him on 6 August 1911. Even when in health he had been less sturdy than his brother Carlos, and more nervous in manner. Unlike Carlos, he expressed himself far more fluently and extensively in writing than in speech, and he rarely if ever addressed an audience except by reading what he had previously written. His writing was clear, and even when at times he apologized for its not being fully polished it was always in good literary style. He was more formal and less gaucholike (*gauchesco*) than Carlos, having spent less of his life in the field and more in learned company. He was handsome, and as he aged he became literally as well as figuratively more highbrowed, although not markedly bald. He had a neat goatee, with a central tuft but wider than a Vandyke beard, and also a large mustache extending beyond and below the sides of his mouth. He wore glasses when reading, but not regularly or when being photographed.

Commentators have sometimes said that Florentino worked in almost complete isolation, which heightened his accomplishment and excused his errors. It is true that there were few other paleontologists in Argentina during his lifetime, and almost without exception he was at outs with those few, including Burmeister, Moreno, and Roth. Nevertheless, it is clear that he was familiar with most of the relevant studies of paleontologists and zoologists elsewhere in the world, and he corresponded extensively with many of them and with other scholars. With this broad and continuing background, it is not really true that he was professionally isolated.

He was enormously prolific, his collected published works on vertebrate paleontology running to seventeen volumes, with another volume of posthumous publications and of manuscripts published only in the collected works long after his death. His collected scientific correspondence fills four fat volumes. As early as 1889, before many of the major collections made by Carlos Ameghino, Florentino published a tremendous work, *Contributions to the knowledge of the fossil mammals of the Argentine Republic* (in Spanish), running to well over a thousand pages and accompanied by a separate volume in large format with 98 plates of illustrations. Some other publications have already been mentioned, and in particular his 1906 book on "The Sedimentary Formations . . ." will be noted later in this chapter.

It is probable that Florentino named as new more genera and species of extinct vertebrates, mostly mammals, than the total of extinct and living species ever so named by any other one zoologist or paleontologist in history. As noted in Chapter 4, description of the first major collection made by Carlos in the Santa Cruz beds of Patagonia were placed by Florentino in 83 genera and 144 species, almost all of them named as new. When Carlos collected in other beds, both older and younger, these numbers were greatly increased, and it has been estimated that in his active lifetime, approximately a short thirty-five years, Florentino named over a thousand species of extinct mammals.

When Carlos began sending him mammals from the older Patagonian faunas Florentino had already used up so many generic names, which cannot validly be used for more than one genus of animals, that he resorted to a dodge possibly never used before and only rarely since. He combined the given names and surnames of other scientists into Neo-Latin or Pseudo-Latin to make new generic names. This gives an interesting glimpse into Florentino's scientific background, as the names thus used were of men whom he knew by their work or in many cases personally. Among these are the following, arranged by nationality:

French—
- *Albertogaudrya,* for Albert Gaudry, 1827–1908
- *Amilnedwardsia,* with an initial as well as a given name for A. Milne-Edwards, 1835–1900
- *Edvardotrouessartia,* for Edouard Trouessart, 1842–1927
- *Henricofilholia,* for Henri Filhol, 1843–1902
- *Paulogervaisia,* for Paul Gervais, 1816–1879
- *Victorlemoinea,* for Victor Lemoine, 1837–1897

German—
- *Carolozittelia,* for Karl Alfred von Zittel, 1839–1904
- *Ernestohaeckelia,* for Ernst Haeckel, 1834–1919
- *Maxschlosseria,* for Max Schlosser, 1854–1932

English—
- *Asmithwoodwardia,* another with an incorporated initial, for A. Smith Woodward, 1864–1944
- *Carolodarwinia,* for Charles Darwin, 1809–1882

Guilielmofloweria, for William Henry Flower, 1831–1899
Oldfieldthomasia, for Oldfield Thomas, 1858–1929
Ricardolydekkeria, for Richard Lydekker, 1849–1915
Ricardowenia, for Richard Owen, 1804–1892
Thomashuxleya, for Thomas Henry Huxley, 1825–1895

North American—
Edvardocopeia, for Edward Drinker Cope, 1840–1897
Henricosbornia, for Henry Fairfield Osborn, 1857–1930
Josepholeidya, for Joseph Leidy, 1823–1891
Othnielmarshia, for Othniel Charles Marsh, 1831–1899

Except for Charles Darwin, whose work in Argentina long antedated the Ameghino brothers and whose life only marginally overlapped Florentino's, those names include virtually all the eminent paleomammalogists contemporaneous with Florentino. Among South Americans the only one whom Florentino honored in this way was his brother Carlos, whom he made the eponym of *Caroloameghinia*. It is not forbidden by the rules, but is considered poor taste, to name a genus for oneself. I honored Florentino posthumously by adding him to his otherwise nearly unique list, naming a quite peculiar early mammal found in Patagonia *Florentoameghinia*.

Having here given some stress to the extraordinary number of genera and species of fossil mammals named by Florentino Ameghino, I must now somewhat reluctantly state that he named far too many. "Reluctantly" because Florentino is an idol of mine, as of many of his fellow countrymen, and yet it has been my lot to see that our idol had feet of clay, metaphorically speaking. His fault was that he did not think in terms of populations rather than individuals, and he never considered the fact, clearly stated by *his* idol, Darwin, that specific populations always have considerable individual variation and that this is essential for the origin and evolution of species. It must also be taken into consideration that Florentino worked rapidly and under intense self-generated pressure, not always giving himself time to make all the needed comparisons of the specimens that Carlos was so plentifully sending in from the field.

Restudy of the Ameghino collection with additional specimens from what he called the *Notostylops* fauna shows that he placed in three different orders (Prosimiae, Condylarthra, and Perissodactyla, for those especially interested in such matters), four different families, seven different genera, and seventeen different species, all named as new by him but all now understood as belonging to just one species (*Henricosbornia lophodonta*). That is an extreme example, but it illustrates the weakness of Florentino's typological and not populational approach to classification which led him to consider as new every specimen that differed in any detail from those with which he compared it. In Ameghino's favor, or at least an excuse for him, is the fact that many zoologists and paleontologists in that period tended to share his typological approach and had not learned the extent of variation in specific populations or how to recognize and measure this.

Somewhat more serious and harder to explain is the fact that Florentino greatly overestimated the ages of all the faunas that he studied except a few of the most newly recent ones. What can only be called a bias regarding the antiquity of the mammalian faunas known to him can be observed in publications as early as 1883 and 1885, when collections made by Pedro Scalabrini in northern Argentina were studied by Florentino Ameghino and on hardly any sound evidence were considered by him to be Oligocene in age. As will be more fully considered in Chapter 7, the North American paleontologist J. B. Hatcher spent three field seasons in Patagonia in 1896–1899. He made geological and geographical explorations in central Santa Cruz province, and he made a large collection of fossil mammals almost entirely from the Santa Cruz formation and of the type Santacrucian fauna. An unfortunate result, which puts both combatants in a bad light and seems in retrospect to have been unnecessary, was one of the most bitter feuds in which Florentino was involved.

In 1896 Florentino Ameghino published a 33-page paper with the title (in Spanish) "Notes on Questions of Argentine Geology and Paleontology." A somewhat altered English version was published in the *British Geological Magazine* in January 1897. In November 1897, after his first Patagonian field season, Hatcher published in the *American Journal of Science* a 28-page paper entitled "On the Geology of Southern Patagonia." In this he differed from Florentino Ameghino, but still rather mildly, especially about the age of the Ameghinos' "Pyrotherium beds," as follows:

> I seriously question the stratigraphic position of the Pyrotherium beds as determined by the brothers Ameghino, although it may seem presumptuous on my part, since I was unable to identify the beds at all, and the explorations, travels and opportunities for observations in this region of Señor Carlos Ameghino have been far more extensive than have my own.

Florentino had considered this fauna to be late Cretaceous in age, but Hatcher argued, incorrectly but on available evidence not unreasonably, that they were much later in age and possibly younger than the Santa Cruz beds. In 1900, when his Patagonian collecting was over and study of the collections was under way, Hatcher published another paper in the *American Journal of Science* on "Sedimentary Rocks of Southern Patagonia." Here he omitted the "Pyrotherium beds" in his table of Patagonian sedimentary strata, writing in his text, "The Pyrotherium beds, as that term has been used by Dr. Ameghino, includes a series of deposits of varying age from Eocene to Pleistocene." Hatcher also asserted that Carlos Ameghino did not determine that the Santa Cruz beds were above the Patagonia formation until his fifth field season, whereas Carlos did determine this on his first field trip and always confirmed it thereafter (see Chapter 4). Hatcher also averred that Carlos (or "Charles" as Hatcher sometimes called him) had changed his mind about whether the Pyrotherium beds were below or above the Patagonia marine formation. In actuality, (as also noted in Chapter 4) when Carlos discovered the Pyrotherium fauna in Patagonia he clearly stated that it was below those marine beds, and he always (correctly) continued so to affirm. All in all, it must be admitted that Hatcher was in serious error and was misrepresenting the Ameghinos in several respects.

In keeping with his character it was inevitable that Florentino riposted in due course (and in French) with "The Age of the Sedimentary Formations of Patagonia," in several parts, totaling well over two hundred pages in the *Anales de la Sociedad Científica de Argentia*. This was devoted in largest part to counterattack on Hatcher. Hatcher had complained that he could not check on the Ameghinos' work because they would not give him the localities where Carlos had collected. On this Florentino said:

> The researches of C. Ameghino have been made at my expense, for my personal instruction, and at the profit of science. I do not have the obligation to make detailed reports. I have always held to giving resumés of the geological researches of my brother to the end that one could take note of the succession of the faunas that I was describing; but I have not had the intention, and I did not either have the obligation, to bring out a guide with the necessary instructions for the harvesting of fossils.

I hasten to tell the reader that this attitude (even had it been exactly true on Florentino's part) has its only really close parallel in the famous Marsh-Cope feud in nineteenth-century United States. It does not now exist among paleontologists and has long been considered unethical.

Hatcher's "review" was highly personal to the point of becoming sarcastic:

> Even so capable a man as Dr. Ameghino should, I believe, find some difficulty in carefully determining the exact sequence of strata in Patagonia from the window of his study, situated in La Plata or Buenos Aires.

Having lost his temper, as "Dr. Ameghino" had also, Hatcher here ignores the fact that the field observations of the "sequence of strata in Patagonia" were made by Don Carlos Ameghino. These have proven correct in almost all instances. It was not the sequence but the ages applied to its members in the geological time scale that were the work of "Dr. Ameghino" and that have in most instances been found to be incorrect.

The most important new evidence on the ages of the Patagonian faunas, as Hatcher pointed out while getting the sequence mixed, was the study by Dr. Arnold Ortmann of the invertebrate fossils from the marine Patagonia formation. Ortmann found that these fossils, unlike the land mammals, resembled marine fossils found on other continents. That resemblance strongly indicated that this thick formation approximately straddled the line between late Oligocene and early Miocene, that is, by still later measurements, close to 25,000,000 years before the present. Thus the Ameghinos' Colpodon fauna, which immediately underlies the Patagonia beds and was overlooked by Hatcher, must be at least close to late Oligocene, and the Santa Cruz beds, just overlying the Patagonia formation, must be at least close to early Miocene. Thus the Hatcher-Ameghino feud should end with both close to right, the Ameghinos as to sequence and Hatcher as to ages, and both definitely wrong, the Ameghinos as to ages and Hatcher in several respects as to sequence.

With Hatcher permanently back in the United States and otherwise engaged, Ameghino might have left it at that. It was, however, inevitable that someone else would differ with Ameghino's dating, even then considered wildly improbable by most non-Argentine geologists and paleontologists. This next happened in 1905, when a German geologist, Otto Wilckens published a paper on "The Marine Beds of the Cretaceous and Tertiary in Patagonia." This long paper—almost a hundred pages—was not confined to the marine deposits but discussed also the dating of the beds with land mammals. While praising the accomplishment of the Ameghino brothers in collecting and naming those mammals, Wilckens differed radically with Florentino's dating of them. He said that he had long hesitated because "[Florentino] Ameghino answered any criticism with the most violent assaults" ("mit der heftigsten Angriffen").

Florentino was busily writing a monograph on the fishes of the sedimentary deposits in Patagonia when he received this blast from Wilckens. He immediately counterattacked with a book (in French) of 568 printed pages, with three large folding plates and 358 figures in the text. On page 2 he quoted Wilckens's remark about "violent assaults" and responded: "Mr. Wilckens has been mistaken in believing that his criticisms could irritate me; quite the contrary, because it is quite agreeable to me to have occasion to expound the facts such as they present themselves. It is not with my views that he should be in agreement but with the facts correctly interpreted."

He then proceeded at enormous length to present the "facts" as *he* interpreted them. This book was published in 1906 under the prolix title (here translated from the French) "The Sedimentary Formations of the Upper Cretaceous and of the Tertiary of Patagonia with a Parallel between the Mammalian faunas and Those of the Old Continent"—or "Old World" in present usage, although some of his comparisons or "parallels" did also bear on North America.

Florentino complained that his many critics kept him so busy replying to them that he had too little time for more important work. This book, more simply referred to as *Les Formations Sédimentaires,* is a reply to criticism, but it is also the most instructive and for many purposes the most useful of his almost two hundred publications. It sums up almost all the combined work of the Ameghino brothers, Carlos the observer and collector in the field, Florentino the interpreter and systematist at his desk. Hatcher had some reason to complain that the Ameghinos did not tell him exactly where their specimens were found, although Florentino had said that some crucial fossils came from the vicinity of Lake Colhué-Huapí. If Hatcher had bothered to go to that region he could not have failed to see the Great Barranca and to have found, among other things, what the Ameghinos called the Pyrotherium fauna, which Hatcher failed to find elsewhere and which he consistently misplaced in the sequence.

In *Les Formations Sédimentaires* are maps by Carlos, somewhat crude but adequate under the circumstances, of the known outcrops of all the mammal-bearing deposits from the "Notostylops beds" below to the "Santa Cruz beds" above. Here also are cross-sections of the rocks at key localities, and in folded Plate III a tremendously long geological profile by Carlos of the region on or near the Patagonian coast all the way from the mouth of the Río Negro in the north to Cape Virgenes in the south. Here, too, are descriptions, discussions, figures of specimens, and in some cases phylogenies (diagrammatic lines of evolutionary descent) for many of the groups of fossil mammals found in Patagonia. Also extremely useful are the complete faunal lists of orders, families, and genera (descending categories in Linnaean classification) for all the faunas recognized as distinct by Florentino, not only in Patagonia but anywhere in Argentina, from the oldest then known—the *Notostylops* fauna or here Notostylopéenne—up to and including the now living fauna.

Anyone following admiringly but cautiously in the footsteps of both Carlos and Florentino Ameghino, as I have, finds that Florentino's recording and interpretation had from the start and throughout his life three major biases or *partis pris.* First, as previously exemplified and least important of the three, when an individual differed at all in comparison with others he considered it to be new (which was rarely true at high taxonomic levels) and gave it a new specific or generic name. As also previously noted he did not perceive that most of the ungulates he described and named belonged to a new order, the Notoungulata of Santiago Roth.

Second, as also previously mentioned but requiring some further explanation, he considered nearly all of his pre-Recent faunas as much older than they really were. This was the main point that his critics made, but they made mistakes themselves that enabled Florentino to reject this criticism. It was believed by all geologists of his time, and still is now, that a fauna including dinosaurs is not younger than the Cretaceous period of the geological time scale, by modern measurement therefore older than approximately 65,000,000 years. Florentino interpreted his Pyrotherium, Astraponotus, and Notostylops faunas as Cretaceous because he thought they were all contemporaneous with dinosaurs. This was not based on reliable field observations because Carlos Ameghino never found dinosaurs associated with these faunas or at a higher level in the sequence. For reasons to be given next, Florentino wanted these faunas to be Cretaceous in age and so accepted unreliable or even false data from others than Carlos (for example Santiago Roth)—data that might just possibly, but as it turned out fallaciously, be correlated with (considered of the same age as) other beds that did contain dinosaurs. In the 1906 book Florentino announced that Carlos had by that time found remains of dinosaurs in exactly the same beds as the "*Notostylops* fauna." He called these specimens "debris of dinosaurs" and "imperfectly known." In fact they were only isolated teeth, and on much later reexamination it was clear that they were *not* dinosaurian. Florentino's jump to that conclusion was only wish fulfillment. In 1931 Carlos agreed with that view and reaffirmed that he had never found an identifiable fragment of a dinosaur in the mammal-bearing beds that Florentino called Cretaceous. No one since then ever has either.

As has been shown here, the Ameghinos called the successive, distinguishable strata and faunas sometimes by the name of a mammal found in them, e.g. "*Notostylops*" or "Notostylopéen," but sometimes by a geographic name, e.g. "Santa Cruz" or "Santacrucien." Starting with Gaudry in 1933 (see Chapter 6) and developing ever since then, these beds and faunas have been given geographic or type-locality names ending in *-ense* or *-iense* in Spanish and now *-an* or *-ian* in English. Ameghino's usage in 1906 and a version of modern usage are given in the following table. It is clear that Florentino inserted into the sequence ages that are now not considered separable and clearly defined at that level. He also inserted some that have turned out to be spurious or misapplied, notably the "Téquéen," "Pehuenchéen," "Protéodidelphéen," and "Tardéen." There is a space in the present system below the Riochican, with a lacuna unknown to the Ameghinos, where it seems that "Protéodidelphéen" and "Tardéen" might fit in, but both those supposed mammalian faunas are spurious. The former was based on scraps of undeterminable age but not belonging here in the sequence and the latter was based on some dubious bits not really identified but almost certainly not mammals.

Now we come at last to the third and worst of the failings of the Ameghinian system, partly based on and strongly interacting with the second, the back-dating of faunas. This failing, although here listed as third, was already involved in the *Filogenia,* published twenty-two years before *Les Formations Sédimentaires*. The latter applies principles of classification and of evolu-

Florentino Ameghino's 1906 Arrangement		Present (1983) Arrangement		
Ages (in French)	Assigned Epoch or Period (in English)	Land Mammal Ages	Epochs	Million years before present
Aimaréen	Recent	Lujanian Ensenadan Uquian	Recent	
				— 0
Platéen Lujanéen	Pleistocene		Pleistocene	
		Chapadmalalan		
Bonairéen Ensénadéen Puelchéen	Pliocene	Montehermosan	Pliocene	
				— 5
		Huayquerian		
Hermoséen Araucanéen Rionegréen	Miocene	Chasicoan		— 10
Mesopotaméen Paranéen [hiatus] Friaséen Magellanéen	Oligocene	Friasian	Miocene	— 15
Santacruzéen Notohippidéen		Santacrucian		— 20
[hiatus] Astrapothericuléen [hiatus]	Eocene	Colhuehuapian		— 25
Colpodonéen Téquéen		[hiatus]	Oligocene	— 30
Pyrothéréen		Deseadan		— 35
[hiatus]		[hiatus]		— 40
Astraponotéen	Upper Cretaceous	Mustersan	Eocene	— 45
Notostylopéen		Casamayoran		— 50
Pehuenchéen		Riochican		— 55
Proteodidelphéen		Itaboraian	Paleocene	
[hiatus]				— 60
Tardéen		[hiatus]		— 65
	Lower Cretaceous		Late Cretaceous	

tionary descent in the *Filogenia* to the fossil record of mammals in Argentina as known to the Ameghinos. As early as 1831 Henri Marie Ducrotay de Blainville (1777–1850) had published a general classification of living mammals. Blainville had worked under and with Cuvier, although they later became bitter enemies and although both were Lamarck's successors as professors of zoology in the Paris natural history museum. Blainville's classification divided the class of (living) mammals (Mammalia) into three subclasses: "monodelphes," now generally called "placentals" (Placentalia or Eutheria), the great majority of living mammals including mankind; "didelphes," now generally called "marsupials" (Marsupialia or Metatheria), including most of the living Australian mammals and the opossums in the Americas; and "ornithodelphes," now called "monotremes" (Monotremata or Prototheria), including among living mammals only the Australian platypus (*Ornithorhynchus*), and the two genera of echidnas or spiny anteaters (*Tachyglossus* in Australia, *Zaglossus* in New Guinea, the latter unknown to Blainville.)

By the time (1884) that Florentino's *Filogenia* was published, that triple subdivision of the living mammals was well established, with some variation in nomenclature. Noteworthy examples include the 1870 classification by the American Theodore Nicholas Gill (1837–1914), that of 1872 by the British Thomas Henry Huxley (1825–1895), and that of 1883 by the British Sir William Henry Flower (1831–1899). By the turn of the century that basic arrangement of mammals, not only the living ones but also those of the whole of the Cenozoic (the Age of Mammals) had been adopted by all mammalogists—except Florentino Ameghino. (Some further complication not affecting any known Cenozoic mammals has occurred since then.) In the *Filogenia* Florentino Ameghino noted quite correctly that his old faunas included mammals different from most of those on other continents, and he believed incorrectly that they were older than those even analogous to them elsewhere. He also decided, incorrectly according to everyone else then and now, that the marsupials were not a distinct, natural group but included separate ancestors of placental groups that he believed to be later in age.

Florentino did not publish a detailed classification of all then known mammals in the *Filogenia,* the *Formations Sédimentaires,* or elsewhere. In the later (1906) book he did however discuss the relationship and classification of some main groups and illustrated supposed phylogenies for several of them. He started with what he called "Protongulés," a term meaning "first ungulates or hoofed mammals." These comprised, in supposed ancestral-descendant sequence, *Proteodidelphys–Caroloameghinia–Pleuraspidotherium.* The first two are marsupials, as he noted. The last is a placental ungulate from the early Cenozoic (Paleocene) of France. *Proteodidelphys* is of completely unknown age but certainly not lower Cretaceous as believed by Florentino. *Carolomeghinia* is also a marsupial, opossumlike but on a sideline without known descendants. Its age is now known to be early Eocene. *Pleuraspidotherium* is thus *older* than its supposed ancestor. Florentino concluded that this sequence "proves not only that the [placental] ungulates are the direct descendants of polyprotodont [relatively primitive] marsupials, but also that this transformation [i.e., evolution] took place on the ancient Patagonian continent." These "protongulés" are supposed to have reached Europe somehow but not North America.

Such non sequiturs go on, one group after another: according to Florentino, *Hyracotherium* (the technically correct name of "eohippus") is not ancestral to horses; the "Notohippidae," a group of South American notoungulates, are. Horses, tapirs, and rhinoceroses do not belong in the same order (Perissodactyla), as everyone else has it, but are of "completely distinct origins"—in Patagonia, of course. The elephants and mastodons evolved from ancestral Patagonian pyrotheres. Florentino formerly thought that artiodactyls—the even-toed ungulates such as cattle, antelopes, and many others—arose in the Old World but from a more primitive Patagonian migrant. In 1906 he decided that they did themselves evolve in Patagonia and spread from there all over the world except Australia. The (placental) carnivores of the rest of the world evolved from a Patagonian group that Ameghino called "Sparassodonta." The fact that these are quite clearly marsupials did not for him make any difference to their being ancestral to all the placental carnivores. All the rodents, he believed, evolved from Patagonian marsupials called Polydolopidae.

Without more elaboration of this thesis, it can be said that for Florentino Ameghino all the mammals in the world had as their ultimate ancestors Patagonian marsupials of various sorts. This was an unfortunate, unbreakable obsession. All that need be added here is that according to Ameghino man also arose, if not in Patagonia, certainly in Argentina—but there from Patagonian ancestry among primitive marsupials allied to the living opossums. That is clearly and fully set out in *Les Formations Sédimentaires.*

Some of Florentino's evolutionary sequences are acceptably reasonable if considered somewhat broadly and confined to fossil mammals then known within Argentina, but virtually all go wrong when projected into lineages elsewhere, and few are just wrong from beginning to end. That is notably the case with his early and late views on the antiquity and ancestry of mankind. His obsession for this topic is evident in much of his earliest work, especially 1875–1879. Further pursuit of it was largely put off while Carlos's collections from Patagonia were pouring in. It emerged at full strength again in the compilation of 1906, in *Les Formations Sédimentaires.* From then on for the short rest of his life most of his attention and writing were on this subject, including all three of his posthumous publications in 1912.

I have brooded a long time on the reasons for this ineradicable bias, this idée fixe, and I have not found an answer. Perhaps my treatment of the subject here will seem unduly critical. Certainly it would have seemed so to Florentino Ameghino, although it did not seem so to his brother Carlos when I discussed this with him twenty years after Florentino's death.

I think that the many people who have collected and studied fossil mammals from South America since Florentino Ameghino would agree with me that he was a worthy precursor. He had his flaws, as who does not? Remember that Darwin believed in the inheritance of acquired characters and proposed a definitely false hypothesis to explain that incorrect view.

I close the discussion of the lives and work of the Ameghinos with a sentence from the introduction to my monograph *The Beginning of the Age of Mammals in South America:* "The partnership of the Ameghino brothers was an outstanding example of teamwork, and their achievement was one of the most remarkable in scientific history."

Notes

The voluminous collection of the life, works, and scientific correspondence of Florentino Ameghino collected and edited by Torcelli is here cited in the notes on Chapter 4. Volume 1 contains a prologue followed by a considerable number of items labeled as biographies. These are not biographical in the usual sense of that word but are rather eulogies spoken or written after Florentino Ameghino's death. They do, however, include many bits of information that have here been culled, collated, and rather briefly summarized. As to what sort of man Florentino was, some of the eulogies include anecdotes which I have generally not repeated because they smack too much of the sorts of petty legends that grow up when a famous character dies. Some of the eulogists speak in glowing terms of Florentino's published works, as is normal in eulogies for writers; it is clear, however, that the eulogists were not just unwilling but were unqualified to analyze or evaluate these works. The correspondence published by Torcelli is enlightening. Some further use is made of it here, but it was especially helpful for Chapter 4 because much about Carlos's movements and work are recorded more in the correspondence between him and Florentino than anywhere else.

Essentially complete lists of the publications by Carlos (few) and by Florentino (a multitude) are given in:

A. S. Romer, N. E. Wright, T. Edinger, and R. Van Frank. 1962. *Bibliography of Fossil Vertebrates Exclusive of North America, 1509–1927.* The Geological Society of America, Memoir 87 (2 volumes). The bibliographies of the Ameghinos are on pages 22–31 of Volume 1, with an additional reference to the joint publication by H. (not P.) Gervais on page 524.

Pages 19–26 of the following publication are a general discussion of the work of both Carlos and Florentino Ameghino. Although I still greatly admire the accomplishments of both the brothers, I now think I was a bit too laudatory and not quite enough analytical when I wrote this long ago.

G. G. Simpson, 1948. *The Beginning of the Age of Mammals in South America.* Part I. Bulletin of the American Museum of Natural History, Vol. 91: Article 1, pp. 1–232.

Tournouër; Gaudry

As will appear, few details of André Tournouër's life are known, but the essential points for the purposes here can be noted. They are important primarily because Tournouër provided the materials for Albert Gaudry's contributions to the study and significance of the history of mammals in South America.

André Tournouër was the son of Jacques-Raoul Tournouër, an artist who was also enthusiastic about paleontology and geology. From 1866 to 1879 the elder Tournouër published some seventeen papers on fossil mammals and on the geology of the rocks in which they had been found in France. In 1878 and 1879 André Tournouër himself had abstracts of two studies of fossil horses published in the bulletin of the Geological Society of France. Thus he was then already following his father's interests.

In 1898, as noted here in Chapter 4, André Tournouër was in Argentina and was undertaking, with limited success, to follow up the Ameghinos' discoveries in Patagonia. Thereafter the Ameghinos, and more particularly Carlos, helped him with information on fossil localities in Patagonia, and he made five expeditions there, ultimately with ample success. How his attention was called to this field was eventually recorded in a French publication by Albert Gaudry in 1906, some years after Tournouër's principal activity in Patagonia. "One day," Gaudry recalled,

> as he [André Tournouër] returned from Mendoza [in Argentina] I talked with him about the discoveries of Monsieur Ameghino and Monsieur Moreno in Patagonia; I asked him to undertake some excavations, in order to emulate his father [Raoul Tournouër] for the honor of French science. At the name of his father he turned toward me a profound and affectionate look: "I will try to emulate him," he said to me; "I am going to go to Patagonia, the Paris museum will have some fossils." He has stoutly kept his word.

It is not quite clear when or where this inspiring suggestion from an elder Gaudry to an active young Tournouër was made. One annotator wrote in 1902 that Gaudry had visited André Tournouër in Argentina, but by 1902 André was already well along in this emulation of his father. At some time not exactly recorded but evidently between 1879 and 1898 André had emigrated to Argentina and was farming near Mendoza, a relatively small but active and prosperous city at the foot of the Andes in north-central Argentina, a long distance from Patagonia.

In the course of his expeditions to Patagonia André ultimately made large collections of fossil mammals and also of fossil marine shells. All of these were shipped to Paris and housed in the paleontological unit set up under the direction of Gaudry in the Jardin des Plantes as a part of the increasingly widespread national museum of natural history. The collections are still there and, as will be further related, the fossil mammals particularly were being studied by Gaudry in his last years and almost to the day of his death in 1908.

After his two brief notes on French fossils, already mentioned, there came André's first publications with some relationship to South American mammals. Published in 1900 and 1901, these were brief comments on "*Neomylodon,*" a mythical rumored Patagonian living ground sloth.

It is here something of a digression but not inappropriate to review briefly one of the peculiar episodes of history which came to involve the Ameghino brothers, Tournouër, and eventually Gaudry. In 1898 Florentino Ameghino wrote that some years previously one Ramón Lista, since then killed by a fall in the Andes, had told Florentino and Carlos that he had seen but failed to kill a strange, anteater-like animal in Patagonia. A traveler, unnamed, had shown Florentino a scrap of old hide containing bony nodules such as had been found with the skeletons of otherwise extinct ground sloths. In 1899 Florentino wrote more positively on "The Present Survival of the Megatheres of the Ancient Pampa" and quoted extensively a letter from Carlos saying that Tehuelche (southern Patagonian) Indians had described at length an animal that they called "Iemisch" and had related in detail the (extremely improbable:) habits of the animal. Even earlier in that year Florentino had concluded that this semi-mythical animal was a live, surviving ground sloth and had given it the name *Neomylodon listai,* "Neo" for recent or surviving, "mylodon" for a fossil (and extinct) ground sloth discovered by Darwin and named by Owen, and "listai" for the deceased informant Ramón Lista.

This astounding news attracted big game-hunters from far and wide, and drew the attention of many paleontologists, among them both Tournouër and Gaudry. Excitement grew from the report by Einar Lönneberg, leader of a Swedish expedition to Tierra del Fuego, that he had found similar bits of hide in Eberhardt Cave at Last Hope (Ultima Esperanza) inlet in far southern Chile. Further exploration confirmed that this cave did contain fairly fresh, unpetrified, bones, hide, and much dung of a ground sloth closely allied, at least, to *Mylodon* and believed to be associated with indications of human activities.

In 1899 to 1901 Gaudry wrote five brief notes on this discovery. Independently, André Tournouër wrote two papers on the subject. The first, in 1900, was "On the *Neomylodon* and the hyminché of the Tehuelche Indians." His "hyminché" is evidently a derivative of the Ameghinos' "iemische." His second and last paper on this subject (like the first, only two pages long) was published in 1901: "On the *Neomylodon* and the mysterious animal of Patagonia."

Not to drag this out at too great length, let it be here said at once that the Indian tales of the "iemisch" or "hyminché," if not just invented to amuse the stupid white men, were simply myths with no foundation in reality. Careful investigation, by the able North American anthropologist Junius Bird among others, found no acceptable evidence that the ground sloths in the Eberhardt Cave had been associated with Indians, even very early prehistoric ones. Application of reliable radioactivity dating methods further revealed that the animal matter in the skins and bones is about 10,000 years old or more. In short, the whole episode was much ado about nothing: there are no living ground sloths and have not been for at least one hundred centuries. The once famous *Neomylodon* is a synonym or at most a subgenus of the classical *Mylodon*.

Tournouër collected in the faunas now designated in English as Casamayoran, Deseadan, Colhuehuapian, Santacrucian, and Pampean (the latter as a collective for the Pleistocene as a whole). As is visible in the table on page 90, there are several ages and faunas between Santacrucian and Pampean. These are not represented in the parts of Patagonia where Tournouër collected, and also were not well known at first hand to the Ameghinos, some of them being from the westernmost part of Patagonia, toward the slope of the Andes, and others from parts of Argentina well north of Patagonia and of the most fossiliferous parts of the pampa in Buenos Aires Province. Paleontological exploration of these areas will be more particularly treated in later chapters.

Tournouër had collected on or near the Great Barranca, mainly in the *Colpodon* fauna of the Ameghinos, subsequently called Colhuehuapian. He had also collected extensively from the Ameghinos' *Pyrotherium* fauna now designated as the Deseadan beds, age, and fauna. Early in 1903, when he had met Carlos and Florentino at Cabo Blanco, he went on along the coast to the vicinity of Punta Casamayor and made a rather small but quite varied collection of the *Notostylops* fauna in a cañadón or (usually) dry watercourse near but not at Punta Casamayor. Carlos Ameghino had visited this area and had identified beds that he considered of *Notostylops* age, but he collected there no surely identifiable mammals. He told Tournouër that the greatest part of the *Notostylops* fauna was found in the Great Barranca or at Cerro Negro, which is essentially a continuation of the Great Barranca westward and still south of Lake Colhué-Huapí. Tournouër returned there on his next Patagonian expedition later in 1903 and this time did collect some members of the *Notostylops* fauna. However, it appears that he again missed the Ameghinos' *Astraponotus* beds, which are present and fossiliferous in that area, although the best known representation of this fauna was in an area north of Lakes Colhué-Huapí and Musters.

(The fact that Tournouër made parts of two separate field expeditions in 1903 seems somewhat puzzling to those who usually collect in the Northern Hemisphere. In the Southern Hemisphere January and December in the same year are of course both summer months, with winter intervening. July and August in Patagonia are so cold and stormy as to make collecting difficult or impossible. In 1903, between the two fields seasons, Tournouër visited France and then returned to Argentina.)

In 1902 and 1903 Tournouër published in France two notes on the geology and palenotology of Patagonia. He treated the *Pyrotherium* beds with some reserve because the description of that fauna was left to Gaudry, who had not yet studied it in detail. In 1905 Tournouër did publish briefly on the front feet of *Astrapotherium,* one of the most striking members of that fauna. (In the same year Gaudry published "observations" on the same subject and specimens.) Finally in 1914 and 1922, after the deaths of both Florentino Ameghino and Gaudry and long after Tournouër had left Patagonia, Tournouër published again briefly on geology and paleontology in Patagonia. These final notes, still in French, were published in the *Bulletin mensuel de la Société linnéenne de la Seine-Maritime.* I have not seen these last papers by Tournouër, published at Le Havre in an obscure regional publication, and I have no further direct information on Tournouër's last years. It is apparent, however, that he had returned to France, where he died sometime in the first part of the twentieth century.

The richest and on the whole most important part of Tournouër's collection was that of the Ameghinos' *Pyrotherium* fauna. Carlos Ameghino's three main localities for that fauna are along the Rio Deseado, in the Great Barranca south of Lake Colhué-Huapí, and on a large hill now well known as Cabeza Blanca ("white head, or hill") in the valley of the Rio Chico del Chubut. Most, and probably all, of these were also explored by Tournouër.

Here again something of a digression is in order, to pick up a story alluded to in earlier chapters. In 1911 Frederic B. Loomis, a professor at Amherst College, with two students and an experienced camp man from Wyoming, made an expedition to Patagonia funded by classmates of Loomis. They wandered widely, but almost all of their fossil mammals were found at Cabeza Blanca, which Loomis believed, correctly, to be one of Carlos Ameghino's localities for his *Pyrotherium* fauna. Their most important discovery was a skull of *Pyrotherium* itself, still the most nearly complete known skull of that genus. After returning to the United States, Loomis published two books, one (1913) a narrative of the expedition's travels and the other (1914) more technical, with some discussion of stratigraphy and a longer attempt to revise the *Pyrotherium* fauna based in part on copies of figures by Florentino Ameghino and in part on the finds of Loomis's party at Cabeza Blanca.

Following Gaudry, Loomis called the beds in which the *Pyrotherium* fauna occurs the Deseado Formation. The more technical of his two volumes was therefore named *The Deseado Formation of Patagonia*. As Loomis was a sincere, industrious, and likable man, it is unfortunate to have to record that this book is so replete with dubieties and with downright errors, both as to stratigraphy and as to paleontology, that it cannot be considered a real contribution to South American geology or paleontology. His one truly important discovery was the skull of *Pyrotherium,* but even here subsequent study (especially by Bryan Patterson; see Chapter 10) has shown that Loomis's description was inadequate and partly inaccurate and his conclusions definitely wrong. On the basis of much more fragmentary specimens, and with his tendency to find the origins of virtually all groups of mammals in Argentina, Florentino Ameghino had considered *Pyrotherium* a proboscidean, that is, a fairly close relative of elephants and mastodonts. Gaudry, however, rejected that view and considered *Pyrotherium* a special offshoot from primitive hoofed mammals in South America itself. In his book on the Deseado fauna Loomis considered the hitherto unknown skull supportive evidence for Ameghino's view. Later (in 1921) Loomis changed his opinion to one even less likely than Florentino's view and his own (Loomis's) in 1914. He had then decided that *Pyrotherium's* nearest relative was *Diprotodon,* a large extinct marsupial known only from Australia, "making it [*Pyrotherium*] a marsupial instead of an elephant, as I [Loomis] earlier thought." Various other attempts have been made to link pyrotheres with mammals of other continents. In this mélée, as it may well be called, Gaudry can be designated as the early winner in a posthumous memoir published in 1909, which will be further mentioned below. If not beyond doubt, it is at least highly probable that pyrotheres arose as such from more primitive hoofed mammals and only in South America.

One further oddity of Loomis may be allowed mention here: he almost always referred to Tournouër as Tournier. The Tourniers were a French family not related to the Tournouërs as far as is recorded. At least one of them, the Abbé Joseph Tournier, was interested in French fossil mammals and prehistory, almost entirely in the Département de l'Ain, more or less between Geneva and Lyons. There is no evidence that he ever visited South America or was concerned with its paleontology.

Returning to the mainstream of the present account, attention here centers on Gaudry, one of the truly great paleontologists of his generation. Urged by Gaudry, Tournouër emulated not so much his father as his friend and at times guide Carlos Ameghino. His discoveries were passed on to Gaudry, who late in his life was reaching new interpretations in the field of Florentino Ameghino, who outlived Gaudry but only by three years. Loomis was following, or it may fairly be said "blundering," in the footsteps of both Carlos and Florentino Ameghino, of André Tournouër, and of Albert Gaudry. Loomis called on Florentino Ameghino, when he visited Buenos Aires and La Plata in 1911, but he found Florentino on his deathbed. Carlos Ameghino, who was still alive and active when Loomis was in Buenos Aires and La Plata, is not even mentioned in Loomis's detailed account of his visits to those cities.

Gaudry may now be more fully put in this increasingly complicated picture. He was born on 16 September 1827 at St. Germain-en-Laye, a city on the left bank of the Seine a few miles downstream from Paris. He was christened Jean Albert, but later dropped the "Jean" and is always referred to as Albert Gaudry. His father, Joseph Gaudry, was a noted barrister (*avocat*), author of works on jurisprudence, and also a devoted amateur naturalist. He passed on this interest to young Albert, who was destined to make natural history, and more specifically vertebrate paleontology, his profession. While still in school he spent vacations freqenting the quarries on Montmartre, from which Cuvier had obtained fossils of early (Eocene) mammals. Later he spent much time in the Paris museum of natural history. Alcide d'Orbigny, here previously mentioned as having traveled in South America and brought back some fossils for the museum, was married to Albert Gaudry's sister. Albert studied in the museum, especially geology and also comparative anatomy with Ducrotay de Blainville, successor to Cuvier as professor of zoology in the museum. In 1852, when he was twenty-five years old, Albert Gaudry was made a doctor in natural sciences on the strength of two dissertations, one in geology and one in zoology, a combination that inevitably made him a paleontologist.

One of Gaudry's earliest publications, in 1853, concerned "the Jurassic beds in which have been found jaws of mammals at Stonesfield, near Oxford." (This has a personal interest for the present author because when I studied those jaws in the London and Oxford museums many years later my age was nearly the same as Gaudry's when he wrote about them.)

Over a period of years from 1853 on, before he turned to South American fossil mammals, Gaudry made a series of journeys first to Cyprus and later to Syria, Egypt, and Greece, studying regional geology and making extensive collections for the Paris museum of natural history and the Ministry of Agriculture. Most noteworthy was mapping the geology of Attica and making a tremendous collection of fossil mammals at Pikermi, a locality between Athens and Marathon where an exceptionally rich deposit of fossil mammals had been known for some time but not exploited so extensively as by Gaudry. His large two-volume work on the fossil animals and geology of Attica (published 1862–1869) is still the major publication on the subject.

In 1853 Gaudry's brother-in-law d'Orbigny was made the first professor of paleontology in the museum of natural history, a position in which he was followed in 1861 by the Viscount d'Archiac (1802–1868), whose splendid lay name was Étienne Jules Adolphe Desmier de Saint-Simon. Then in 1871 Edouard Armand Isidore Hippolyte Lartet (1801–1871) served briefly before his death, followed in 1872 by more simply named but not less brilliant or accomplished Albert Gaudry. In 1882 Gaudry was elected to the Académie Française, the apotheosis for French scientists.

Unlike his brother-in-law, and indeed a majority of French biologists and paleontologists in his time, Gaudry was an evolutionist. He was also more of a generalizer and theorizer than most. In 1873, 1883, and 1890 he published three books on "The Sequences of the Animal World in Geological Times." (The word I have translated as "sequences" is *enchainements,* more literally "concatenations.") In 1896 he published another book, "Essay on Philosophical Paleontology," a commentary on the less theoretical books about the fossil sequences in geological time. In this latter work he noted the practical use of fossils in stratigraphy:

> I think that geologists will gladly accept the way of seeing that is expressed in my works, because, if it is true that the geological strata are nothing but stages in the history of the development of beings, the knowledge of these stages of evolution will provide a precious aid for the determination of the ages of the earth.

Gaudry is seen in that quotation as more interested in the facts of evolution and its practical application to stratigraphy than in its causes. Nevertheless, he elsewhere and continually indicated a belief that it is more spiritual than physical causes that have determined the course of evolution. Thus he wrote in his book of 1890:

> Evolution has advanced through the ages like a sovereign whose majestic progress nothing could stop. Vital competition, natural selection, influences of the environment, migrations no doubt have aided it, but its principle has resided in a higher region, too high for us to be able, as at present, to grasp it well.

It is here clear, as in some other passages of Gaudry's works well after the publication of Darwin's *The Origin of Species,* that although a determined evolutionist Gaudry was not only non-Darwinian but also anti-Darwinian. This has been pointed out by more recent French paleontologists, most of whom, if perhaps not quite all, are evolutionists, but many of whom, if perhaps not a majority, are non-Darwinian. It has also been suggested that a certain mystic or at least nonmaterialistic element supposed to be involved in evolution (or *transformation,* as it has often been called in French) was handed down through the years from Lamarck to Gaudry, to Gaudry's student Pierre Marcellin Boule (1861–1942), to Boule's student Pierre Teilhard de Chardin (1881–1955), and from Teilhard to a group who may be called cultists.

Darwin and Gaudry corresponded with each other and read some of each other's works. In a letter to Gaudry written on 21 January 1868, Darwin remarked, "How strange it is that the country which gave birth to Buffon, the elder Geoffroy, and especially to Lamarck, should now cling so pertinaciously to the belief that species are immutable creations." (The "elder Geoffroy" referred to by Darwin was the older of the two Geoffroys well-known as zoologists. Étienne Geoffroy Saint-Hilaire [1772–1844] maintained that under changing circumstances species might be somewhat mutable, a sort of approach, but a limited and cautious one, to the idea of evolution. His son Isidore [1805–1861] held a similar opinion.)

We now have the setting in which the studies by Gaudry in his final years were to become pivotal. In 1902 Gaudry published notes on the work of Tournouër in Patagonia almost simultaneously in journals of the Academy of Science in Paris, of the Geological Society of France, and of the Autun (Bourgogne) Natural History Society, of which he was the honorary president. It was reported that Gaudry made his announcement there in person on 29 June 1902 on his return from a visit with Tournouër. (There seems to be some peculiarity in the dating of Gaudry's first announcement of Tournouër's work in Patagonia since we know from a letter by Carlos Ameghino that Tournouër was already seeking fossils in Patagonia as early as 1898.)

In 1903 Gaudry published briefly in the bulletin of the Société Géologique of France on "The advance [*marche*] of evolution in Patagonia," and in 1904 he started publications under the general title of *Fossiles de Patagonie* which continued to, and indeed a year after, his death. The first was "On the dentition of some mammals," a summary of a considerable part of the Tournouër collection with numerous figures. Second in this series in 1906 was "On the postures [*attitudes*] of some animals," and published also in 1906 was "Study of a part of the Antarctic world." The fourth part of *Fossiles de Patagonie,* "On the economy of nature," was published in 1908. The fifth and last part, "The *Pyrotherium,*" was published posthumously in 1909. Gaudry had died on 27 November 1908.

As Gaudry studied the Patagonian collections, his first reaction bordered on dismay. In his three books on the sequences or concatenations of evolving faunas and one on the philosophy of paleontology, as previously mentioned, he had envisioned evolution as an overall phenomenon, with changes in faunas occuring in similar ways and at the same times over most or all of the earth. This was a practical basis for geological correlation: similarity of faunas indicated similarity, or identity, of age of the fossil-bearing strata. Gaudry's own experience was almost all in Greece and France, and this system worked out well enough for Europe. But not for South America, as he soon learned from the Tournouër collections. In the *Fossiles de Patagonie* he wrote:

> Evolution, we thought, had run its course through the ages without anything stopping it. Now behold that Patagonia teaches us that what has seemed true within the limits of the northern hemisphere ceases to be so in the southern hemisphere and perturbs our belief in the similarity of the course of evolution in the world as a whole.

Yet Gaudry came to see that evolution in the two hemispheres—or in his experience in Europe and in South America—could have the same tendencies and yet run along different lines. This and other studies of the fossil mammals collected by Tournouër and those published by Florentino Ameghino started Gaudry on reorientation of concepts of evolution in several ways.

Gaudry's only really complete and detailed studies of the many genera then known among fossil mammals from South America was the last memoir in the series on *Fossiles de Patagonie,* that on *Pyrotherium.* For that genus Tournouër's collection was particularly noteworthy because it included a number of previously unknown skeletal parts, although still not enough to make a good reconstruction of the whole animal. Even this much showed that *Pyrotherium* exemplified the trends in South America for evolution there to go off on its own lines, contrary to Ameghino's views and some still held by a few later students. Gaudry planned to continue with equally detailed treatment of other South American fossils "if," as he wrote, "God lends me life" ("si Dieu me prête vie"). Unfortunately God did not so will. Gaudry died on 27 November 1908 at the age of eighty-one, and the *Pyrotherium* monograph was his last direct contribution to knowledge.

Gaudry's studies of Patagonian fossil mammals thus were left incomplete, and they constituted only a minor part of his otherwise very extensive paleontological studies. Nevertheless, he did lay a basis for what would become criteria for interpretation especially of the early and middle South American fossil record of mammals and their faunal associates. First and on the whole most important was the finding that the early South American mammals were evolving along lines quite unlike those known on other continents. Thus the idea that these mammals were ancestral or closely related to all others elsewhere—almost sacred doctrine to Florentino Ameghino—was highly improbable for any of them and impossible for these faunas as a whole. It became evident, as it still is to most students of the subject, that South America was an isolated continent for some tens of millions of years.

Gaudry's second major reorientation was no less important, although to some extent it paralleled or was even anticipated by studies being made in North America based largely on J. B. Hatcher's Patagonian collection, as noted in the next two chapters. This was the point, already made here, that Florentino Ameghino greatly overestimated the geological ages of the strata and faunas in the sequential series of Carlos Ameghino and some other mainly Argentinian geologists.

The third reorientation by Gaudry seems minor in comparison with the first two but nevertheless has had a result important to those dealing with faunal and stratigraphic nomenclature. As noted in Chapters 5 and 6, the Ameghinos used a mixed nomenclature for strata and faunas based in part on the generic names of mammals in those faunas and in part on geographic names of places or land features at or near which the faunas involved had been first (or at least early) found. Gaudry initiated the consistent use of geographic names, but not quite in the orthography now used. Thus, he applied the name "du Deseado" to what Florentino called the *Colpodon* beds, above, and the *Pyrotherium* beds below. It is interesting that in 1899 Carlos Ameghino, in a letter to Florentino, suggested that the highest fauna in the Great Barranca south of Lake Colhué-Huapí be called "Colheuhuapense," but Florentino did not accept this and called it *Colpodon* fauna and beds. As late as 1924 Anselmo Windhausen was still using the Ameghinian generic names suffixed by *-ense,* but beginning essentially with Lucas Kraglievich (see Chapter 9) the use of geographic names with locative suffixes has become general. It is also unfortunate that Gaudry called the *Notostylops* beds of Ameghino "Casamayor." Tournouër did collect fossils from those beds but some distance from Punta Casamayor in what the Ameghinos called Cañadón Tournouër (now Cañadón Lobos). Carlos Ameghino found only one inexactly identifiable specimen in that region, and the Ameghinos' conception of the *Notostylops* beds and fauna was based essentially on specimens from the Great Barranca. It is also to be noted that when Gaudry applied the name Casamayor he had questionably identified but had not closely studied specimens from his "Casamayor." The name Casamayorense in Spanish and Casamayoran in English has now long been in general use so that it would be unwise to rename it. It is further notable that Gaudry did not rename the *Astraponotus* beds so called by the Ameghinos. They are now called Mustersense and Mustersan after the other of the two lakes nearby.

It is further notable that the Ameghinos and other early paleontologists did not clearly distinguish between faunas, the rocks in which they are found, and the geological ages represented by faunas and rocks. Gaudry, too, was not entirely clear in making these distinctions, but some of his studies did help to direct eventual attention to them.

Perhaps the most important claim for the continued fame of Gaudry is that he was the first evolutionary paleontologist, as several French commentators have said. He certainly was more firmly and outspokenly evolutionary than any of his contemporaries, at least those in France, and this merits acclaim. Yet to call him the *first* evolutionary paleontologist may be questionable. Darwin, after all, was a paleontologist, too, among other things.

Notes

Despite help from colleagues in France I have not been able to find any biographical sketch of Tournouër, or even the dates of his birth and death, or a portrait of him. What has been said of him has had to be gleaned from some of his own few, nonautobiographical publications, from mentions of him in published correspondence of the Ameghinos and of Gaudry, and from mention in Gaudry's publications on Patagonian fossils. He first published a paleontological note in 1878, so was presumably then an adult, and his last publication (known to me by title only) was in 1922, so he was presumably then still alive. There is a specific note that he made five collecting expeditions to Patagonia, and these probably dated from 1898 to 1903.

There is considerably more information on Gaudry's life, although not as much as might be expected for a man so highly esteemed and widely honored. I have specially used a 38-page biographical notice published in 1909 by Dr. X. Gillot, then president of Natural History Society of Autun, and a long thesis or dissertation in French by Jean Albert Durand on "The idea of evolution in the work of Albert Gaudry," dated 1975, unpublished but kindly loaned to me by the laboratory of paleontology of the University of Paris.

THE DISCOVERERS

Thomas Jefferson

Caspar Wistar

Georges Cuvier

Francisco Muñiz

Peter Wilhelm Lund

Alcide d'Orbigny

Charles Darwin

Carlos Ameghino

Florentino Ameghino

Santiago Roth

Albert Gaudry

John Bell Hatcher

William Berryman Scott

Barnum Brown

Elmer Riggs

William John Sinclair

Lucas Kraglievich

Ángel Cabrera

Ruben Stirton

Carlos de Paula Couto

Bryan Patterson

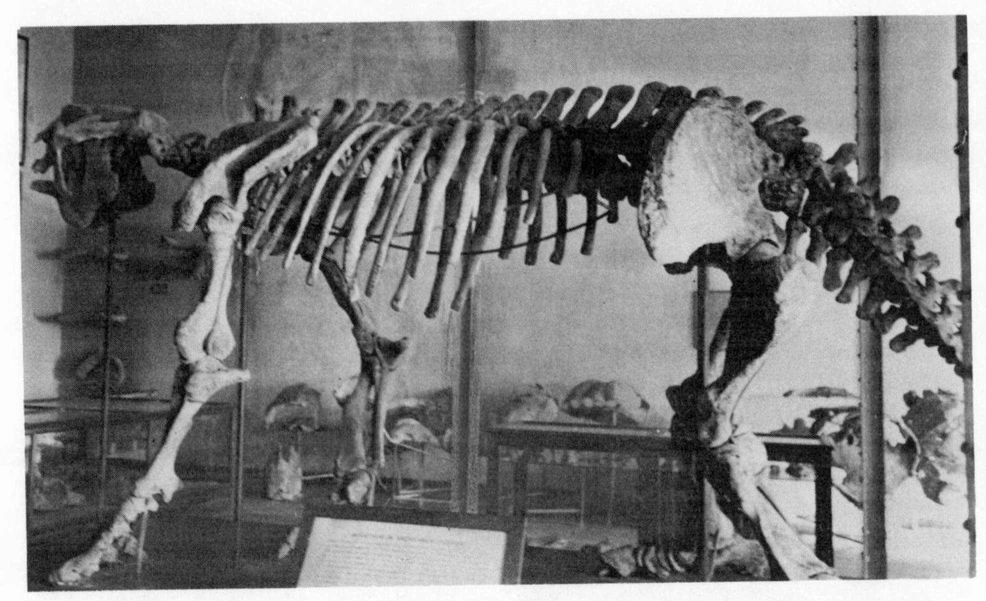

The megathere from Luján as it stands today in the natural history museum in Madrid. Photograph by A. Castellano, courtesy of Bermudo Meléndez, University of Madrid

View of the Great Barranca in Patagonia

Hatcher

The discovery of the richness and the variety of fossil mammals in southern Argentina nearly coincided with equally rich but even more widespread discoveries of quite different fossil mammalian faunas in the United States. Florentino Ameghino's many and detailed publications—mostly in French, an international language—and his displays of fossils in Europe and sale of some of them both to European paleontologists and to at least one North American (Edward Drinker Cope) made knowledge of his riches worldwide. It was inevitable that North American paleontologists would become interested in the Patagonian fossils and eventually would there seek collections for further international studies.

The most active North American nineteenth-century vertebrate paleontologists for some years were Cope, principally at the Philadelphia Academy of Natural Sciences, and Othniel Charles Marsh (1831–1899) at the Peabody Museum (endowed by his uncle) of Yale University. As is still widely known, these two became not merely rivals but bitter enemies who carried on a vendetta until their not widely separate deaths. Both were mainly concerned with North American studies and secondarily with those in Europe; neither visited South America or personally studied fossils from there. As has been noted (Chapter 5), Cope did obtain some South American fossils, and among his almost innumerable publications from 1864 to 1900 (then posthumous) there are five directly concerned with South America: one in 1885 on some Brazilian fossils, three in 1891 and one in 1897 on Argentinian fossils discovered and mostly made known by the Ameghinos. Marsh in 1869, early in his career, did publish one note on fossil reptiles from Brazil, but displayed no special interest in South American fossil mammals. Nevertheless, Florentino Ameghino honored him with the name *Othnielmarshia,* for a small ungulate in the Ameghinos' "Notostylops fauna."

Despite Marsh's lack of personal interest in South America, an important but indirect connection between him and Patagonia did develop. The connection was by way of John Bell Hatcher, who worked for Marsh for some years and soon thereafter became the first North American to make an important collection of Patagonian fossils. To his short but varied and productive life we now turn.

John Bell Hatcher was born on 11 October 1861. The Hatcher family was from Virginia, but in 1861 John and Margaret C. Hatcher, the baby's parents, were living in Cooperstown, Illinois. Soon thereafter they moved to a farm in Greene County, Iowa, where John Bell spent his childhood. When not busy on the farm, the father sometimes taught, and the child received his elementary education in part from his father and in part from public schools. Hatcher was sickly as a child and indeed throughout his life, even though eagerly working, often in difficult and unhealthy circumstances. He was determined to go on to further education and knowledge. As the family was too poor to send him to college he worked as a coal miner until he had saved enough money for that purpose. In the fall of 1880 he entered Grinnell College in Grinnell, Iowa, but soon transferred to Yale College.

His work in the mines had given him a special interest in geology, and also in paleontology, as he had made a collection of fossils, mostly plants, from the coal-bearing beds. The great geologist James Dwight Dana (1813–1895) was then professor of geology and mineralogy at Yale, and Hatcher studied with him. Hatcher and his fossil collection also attracted the interest of Marsh, who was then also a professor at Yale although involved with research rather than teaching. In 1884, not yet quite twenty-three years old, Hatcher was graduated with the degree of Bachelor of Philosophy, then and long thereafter the Yale equivalent of the B.A. degree in most colleges.

Although Marsh had headed several fossil-hunting western expeditions in the 1870s, thereafter he preferred to hire collectors and have the results shipped to him for preparation and study in New Haven. As soon as he had his diploma, Hatcher asked Marsh for a job collecting fossils, anywhere and at any salary. Marsh hired him forthwith and sent him off to Long Island, Kansas, to work collecting fossil bones, there mostly of extinct rhinoceroses, with Charles H. Sternberg, who was already a famous fossil collector and was eventually to be the father of others. Even then, at the start of his professional career, Hatcher was a perfectionist. He loved fossils, and he was determined to collect them as perfectly as possible. He was soon critical of Sternberg and disliked working under his direction. He wrote to Marsh that Sternberg should be "taking more pains in raising the bones after they have been uncovered," and he asked Marsh for permission to "work somewhat independently." Marsh soon did allow him to work on his own or as a head of field parties, and he early became the most successful discoverer of fossil bones and the most skillful excavator of them in his time.

For years Marsh kept Hatcher in the field, sometimes through the winter in spite of the rheumatism that constantly troubled and occasionally immobilized him. At first his collecting was mainly for mammals, but from 1888 it was usually for dinosaurs, notably many specimens of the extraordinary horned dinosaurs collectively known as Ceratopsia and most completely represented by *Triceratops,* three-horned ceratopsian. Hatcher was in charge of this work, but he had at times a fairly large crew, which included C. E. Beecher, who became director of the Yale Peabody Museum in 1894, shortly before Marsh's death. Another member of the party was O. A. Peterson, who later worked also with Hatcher both for Princeton and at the Carnegie Museum in Pittsburgh. In 1887 Hatcher married O. A. Peterson's sister Anna M. Peterson. Of their five children, one died as an infant in 1903, but the others—Earl, Harold, Alice, and John W.—survived their father, as did their mother. After Hatcher's death in 1904 the family moved to Lamont, Iowa.

Before going on to Hatcher's break with Marsh and his transfer to Princeton and expeditions to Patagonia, I should mention the major discovery he made in 1889. While collecting late Cretaceous dinosaurs in Wyoming he found some minute teeth of mammals in the same beds as the dinosaurs. Only one or two scraps of mammals of that age had then ever been found anywhere in the world. (As noted in Chapter 5, Florentino Ameghino considered large numbers of Carlos's finds in Patagonia to be of Cretaceous age, but this was erroneous.) Marsh rushed this discovery of North American truly Cretaceous mammals into print in an 1889 issue of the *American Journal of Science*. (This journal, founded and edited by J. D. Dana, became almost a house organ for the Yale Peabody Museum and especially for Marsh.) Marsh ordered Hatcher to concentrate on the search for more Cretaceous mammals, and Hatcher succeeded in finding and sending them to Marsh in numbers running into thousands, but almost all small, isolated teeth. In 1896, when he was no longer working for Marsh, Hatcher published in the *American Naturalist* (significantly not the *American Journal of Science*) on his localities for Cretaceous mammals and his method of finding so many of them: most of them came from the ant hills so common in the western plains. The ants in excavating their subterranean homes carried out any granular material that they encountered and dumped it on the hill that resulted around the entrance. In many cases this accumulation included fossil teeth and some bits of bone, which were picked over by Hatcher and paleontologists following him. The method has since been applied to collecting other fossil mammal teeth, especially of small rodents, later in age. (More recently the search for small vertebrate fossils has been supplemented or supplanted on a worldwide scale by screen washing. The rock or matrix in which fossils may occur is slacked down, usually in water, and then put through a fine screen that separates the small nodular material, which can be picked over to take out the teeth and other bits of small vertebrate fossils.)

In spite of Hatcher's increasingly brilliant successes in the field, all was not going satisfactorily between him and Marsh. Like some others among Marsh's fossil collectors, Hatcher had difficulty in getting prompt and needed payment for his work and field expenses. He also felt that he could be better employed at the museum during winters. He did not complain at his suffering from rheumatism, but this must also have been involved in his wish for laboratory as well as field work. However, he was so impetuous and so enamored of the great open spaces that he was temperamentally less suited as well as being less experienced for laboratory than for field work.

Matters did not improve between Hatcher and Marsh in 1890–1892, despite the signing of a contract in 1891 promising Hatcher $2,000 a year for five years, with six months a year in field work, five months in the museum at New Haven, and one month for vacation. After a loss of government funding put Marsh in the position of having to pay Hatcher out of his own pocket, the contract was canceled, and Hatcher was in effect discharged in January 1893, with a letter of recommendation.

In 1891 Henry Fairfield Osborn (1857–1935) had offered Hatcher a position at the American Museum of Natural History, but Hatcher had declined this on the basis of his supposedly five-year contract with Marsh. Fortunately for Hatcher—and also, as it turned out, fortunately for the paleontology of South America—on being discharged by Marsh he received an invitation from W. B. Scott to come to Princeton (or "the College of New Jersey at Princeton" as it was then called) as curator of paleontology and assistant in geology. The job involved an immediate expedition into the field. That year and each of the two following, 1894 and 1895, Hatcher took groups of Princeton students out west with him to collect fossil mammals, mainly in areas where Hatcher had collected for Marsh. Most of their finds were excellent in variety and preservation and were well collected in spite of the customarily inadequate funds.

By all reports Hatcher seems to have been good with students and helpful to them, but he still had the urge to get out on his own. He also evidently had the ambition, not clearly expressed, to do more research and publication by himself and not merely to gather teaching and research materials that would be published by others and bring renown to them, as was his function while working for Marsh. An ambitious plan began taking shape, to go to the rich fossil fields of Patagonia already made so famous by the Ameghinos. I have not been able to discover much in the available documents about the origin and development of this plan, but, as will be discussed in the rest of this chapter, Hatcher did make highly successful expeditions to Patagonia.

On the first occasion, he and his brother-in-law O. A. Peterson (who had worked for Osborn at the American Museum in addition to working for Marsh with Hatcher) sailed for Argentina from New York on 29 February 1896. Writing autobiographically about this event, Scott later said that it nearly occasioned "the only misunderstanding that ever arose" between him and his old Princeton classmate Henry Fairfield Osborn. In a sense Scott and Osborn became the followers of Cope and Marsh, with whom their lives overlapped, as leaders in American vertebrate paleontology. Unlike the feuding Cope and Marsh, however, they remained close friends. There will be more about Scott, particularly in the following chapter, but this much is relevant here because it relates to the outset of Hatcher's great expeditions to Patagonia.

As told by Scott in his autobiography Osborn had expected to send an expedition to Patagonia for the American Museum, but he asked Scott to keep this as the strictest secret. Writing years later, Scott said that he had "respected his [Osborn's] confidence and told no one." He added that he had known nothing of Hatcher's plan until it was complete and under way. Otherwise, Osborn might well have thought that his scheme had been passed on to Hatcher, and the offense would have seemed all the worse as Hatcher lured away and took with him O. A. Peterson, who was then employed by Osborn. Scott's explanation was accepted by Osborn, but on other evidence it is improbable that Scott was wholly open and frank in this matter: the fact is, Hatcher had written to Scott from a "Camp on head of Corral Draw" (a locality in the Big Badlands of South Dakota), in a letter dated "March 9, 1894":

> How about that South American trip. What can be done about it! I should like immensely to go & I think it can be done with $1000, or $1500 at the outside. I will give $50. per month toward it from the time I start untill [*sic*] I return if necessary with the understanding that I be remunerated for it within 5 years. By doing this I believe that if I leave Princeton with say $700 it will be sufficient for the trip with my $50. per month. The plan at present is to sail from New York about Aug. 15th. Now I think we aught [*sic*] to take hold of this matter in earnest. Above is my proposition and I am willing to go for 9 months or one or two years. I think it a great chance & will do my best, if it can be carried out.
>
> Let me know as soon as possible how many men there will be on the expedition.

As that was written almost two years before the (two-man) expedition started, Scott was either forgetful or disingenuous if he later assured Osborn that he knew nothing of Hatcher's scheme until it was being carried out.

Even for a man like Hatcher, austere to the point of self-sacrifice, expeditions for fossils were expensive and difficult to finance. Cope and Marsh were both relatively well-to-do, and both also did obtain some limited and temporary help from federally funded geological surveys. At Princeton there was no support for such expeditions. Hatcher's field seasons with students were almost entirely financed by the students themselves or their families; to encourage this Hatcher would generally take the students on a scenic western tour after collecting fossils.

It does not appear that Princeton made any material contribution to the Patagonian expeditions beyond Hatcher's recognition that Scott "gave freely his influence and best efforts toward their accomplishment." However, various "friends and alumni of the University," including some of the students who had traveled with him, or perhaps their families, gave not only moral support for the first two expeditions but also "substantial financial assistance." Hatcher obtained reduced passenger rates to and from South America, and within Argentina he had free transportation on official transports plying to various ports in Patagonia. He later wrote that "the expenses of the third expedition were for the most part met by the present writer," i.e. Hatcher himself. The fifty dollars a month that he was himself prepared to provide for expenses of the expeditions, as noted above, were offered against repayment in five years, but I find no evidence that any repayment was made. In fact, after five years Hatcher was no longer connected with Princeton.

As Hatcher had inherited nothing and surely was not overpaid by Marsh or by Princeton, and as he had a family to support, the affluence indicated by his fifty dollar loan (or contribution) per month and by his personally paying "the most part" of the expenses of the third expedition poses a problem. Whence, indeed, did Hatcher get so much money? The gossiping speculation about this is still sometimes heard even now, long after the deaths of Hatcher and all those directly involved with his work and expeditions. Curiously, this was put into print in a somewhat fantasized form in an article tastelessly called "Patter-gonia" in the *Atlantic Monthly* magazine for September 1937. The tale is that a "professor" from Princeton went to Patagonia for a few months but stayed for three years because he was so skillful at playing poker that he cleaned up on all the moneyed residents. This account is obviously garbled, but the possibility that Hatcher augmented his resources at various times with skill in gambling is not necessarily at odds with his spirited and, in American terms, free western style. (As regards the Patagonian expeditions this story is somewhat oddly included in the chapter on Hatcher in Schuchert and LeVene's biography of Marsh.)

Hatcher made three expeditions to Patagonia, each time with a single companion or field assistant. The most productive stretches of paleontological work in Patagonia coincide with the summer months of the Southern Hemisphere, and Hatcher's three expeditions in 1896, 1897, and 1898 were partly but not accurately aimed for those seasons. (Contrary to the wisecrack in the *Atlantic Monthly,* Hatcher did not go to Patagonia for a few months and stay for three years; he went three separate times.)

The first expedition left New York on the steamship propitiously named "Gallileo" (as Hatcher misspelled the name) either on 29 February 1896 (which was a leap year) or on the following day, 1 March. The party consisted of Hatcher and of his brother-in-law, who in Hatcher's account is first sedately called "Mr. O. A. Peterson" and thereafter only "Mr. Peterson." (By diligent searching, I have been able to find—not from Hatcher's publications or correspondence—that Peterson's "O. A." stood for Olof August.) There were inevitable delays, but Hatcher, Peterson, and their field equipment did finally get off on a government vessel bound for Gallegos, a Patagonian port on the estuary of a river of the same name.

While waiting to embark, Hatcher had seen what sights there were in Buenos Aires and had also visited La Plata, where he saw some of the fossils on display in the museum. However, his account does not mention meeting any of the Argentine paleontologists, which at that time included both Carlos and Florentino Ameghino at the height of their activity.

The transport from Buenos Aires made several stops on its way to Santa Cruz and Gallegos. At Bahía Blanca, Hatcher regretted not having time to visit Punta Alta, where Darwin had collected fossil mammals (as here related in Chapter 2). The first stop in Patagonia itself was at Puerto Madryn, near the mouth of the Río Chubut and the capital of the then territory of Chubut.

Hatcher was not fazed by the discovery that he had timed the expedition to arrive at the beginning of the notoriously severe Patagonian winter. While employed by Marsh he had often camped in the winters of Wyoming, Montana, and the Dakotas, and he did not think that Patagonia could top that experience.

Still, every outlander who has ventured to Patagonia has surely had new impressions and emotions that were never forgotten. Hatcher's account of his Patagonian expeditions was written after his return from the last of them and after he no longer was at Princeton. Published in 1903 in an enormous quarto volume of 314 pages with fifty photographs on inserted plates, two text figures of geological cross sections, and a map of Patagonia and Tierra del Fuego, Hatcher's account is so detailed and sometimes vivid that it must surely have been derived for the most part from a diary or journal written as he went along. Thus it preserves much of his impressions and emotions as they occurred, and only in part and toward the end does it become retrospective.

His descriptions of Patagonia in that first encounter of 1896 are nearly a panegyric:

> We arrived at the entrance to New Bay early one morning, and as we approached, a line of high cliffs rose sheer from the water to a height of perhaps 200 feet and stretched away on either side as far as the eye could reach. This was our first view of that great sea wall that extends

> almost uninterruptedly all along the eastern coast of Patagonia, from the mouth of the Rio Negro to the eastern entrance of the Straits of Magellan, and with which we were shortly to become so familiar. . . .
>
> As we entered the bay, which at its mouth has a breadth of hardly more than a mile, it was seen to expand rapidly into a broad, nearly circular basin. . . . Seen under the perfect atmospheric conditions that then prevailed, it was indeed a thing of beauty. From its surface, smooth as that of a polished table, were reflected images of the high, precipitous cliffs surrounding it. . . . The peculiar beauty of this body of water was emphasized by the dreary, not to say desolate appearance of the surrounding plains . . . which only supported a scanty covering of brown and withered grass.

That evening the captain took the ship farther out in the bay and lowered the boats to take Hatcher and others fishing. That the fish were not biting did not matter to Hatcher:

> . . . For I was chiefly interested in the splendid, if somewhat solitary, beauty of our surroundings, and was overcome and enraptured by the quite unexpected novelty of the situation. From childhood I had thought of the coast of Patagonia as visited by almost perpetual storms. . . . How different our actual experience. . . . The complete quiet, save for the rhythmic murmur of the muffled oars, was comparable only with that described by Arctic travelers.

Hatcher thus encountered Patagonia with feelings close to ecstasy, but Patagonia has many moods, and he was to learn that this was an occasional but not a usual one. The fair weather held only as far as their next stop on the way south, Puerto Deseado, which Hatcher called "Port Desire." He had an erratic way with place names, translating Puerto Deseado to "Port Desire" and, later, Punta Arenas to "Sandy Point," but not calling Santa Cruz "Holy Cross." His goal, Gallegos, also retains that Spanish form in his account. It means literally "people from the province of Galicia in Spain" but in Argentina may be applied, often pejoratively, to anyone from Spain, although Gallegos is also a worthy surname in South America.

Hatcher, Peterson, and their extensive equipment arrived safely in Gallegos on 30 April 1896, just two months after their departure from New York. Here the weather was more typically Patagonian than at the time of Hatcher's first, enraptured encounter with the region; that is to say, now the weather was vile and getting worse. They were hospitably received by the governor of the Province of Santa Cruz, General Edelmiro Mayer, and were guests in his house while in Gallegos. The general was a well-educated, in fact an accomplished man who was fluent in French, Spanish, German, and English, and the house contained a large library, a grand piano, and "a handsome American organ"—but it did not contain easy chairs, fireplaces, any means of heating in this bitterly cold climate, or any of the comforts of the poorest North American house. Hatcher consoled himself by writing, "We well knew that when once we were ready to start and dependent upon our resources, we could, with our equipment and experience in camp life, make ourselves far more comfortable in our tent than here in these cold, damp, cheerless halls."

Before accumulating the necessary horses and a wagon on which to carry their equipment, Hatcher and Peterson went with the governor on an excursion into the interior, with stops at a place called Geir Aike (a single building with one room for the owners, one for paying visitors, and one for sale of liquor and other supplies—a *boliche* in Patagonian terms, although Hatcher did not use that word). They also visited a hacienda of the governor, and a village of Tehuelche Indians. Along the way Hatcher especially noticed a steep river bank or "high sea wall" in which he believed fossils had been found long before. In retrospect he wrote of "the almost embarrassing riches [of fossils] contained in the deposits which [presumably] formed this line of cliffs." This set him to musing in a way that is worth quoting for the insight that it gives into his character and the light it throws on his having successively left employment by O. C. Marsh, spurned the appointment offered by H. F. Osborn, left the employment by W. B. Scott, and finally fretted under the dominance of W. J. Holland for the few remaining years of his life. At this point in his writing retrospectively about a time before he had collected or even seen a fossil in Patagonia, he said:

> The study of nature is always instructive and interesting, even inspiring and impressive, if the student be a real lover of nature seeking for truth at first hand and for truth's sake, and not merely a fireside naturalist, who seldom goes beyond his private study or dooryard, and either contents himself, like other parasites, with what is brought to him, or like a bird of prey forcibly seizes upon the choicest morsels of his confrères, with little or no consideration for the rights or wishes of those who have brought together the material at so great an expense of time and labor.

It does not take much reading between the lines to see who the "real lover of nature" was and who the "parasites" were, and later Hatcher made his point even clearer. He compared the anxieties, uncertainties, and labors of prospectors for precious metals and collectors of vertebrate fossils, and then added, "Indeed the simile may be carried still further, for both are too often deprived of their fair share of the gains that follow, which in the one case take the form of money and in the other that of credit for work done."

Back at Gallegos, Hatcher managed to purchase horses for Peterson and himself. They had brought saddles and other riding gear from the United

States, but they needed, or at least wanted, wheeled transportation for their equipment and provisions. The extremely ponderous carts used with oxen to haul wool to the port were hopeless for their purposes, but they finally did find a relatively light two-wheeled, three-horse cart. At last on 16 May they set out to seek what "choicest morsels" they might find and collect.

They went back to Geir Aike, had some trouble fording the Río Gallegos, found no fossils in the cliffs that had so inspired Hatcher at first sight, so went down the river to Killik Aike, where abundant fossils were reported on the estancia of H. S. Felton. After some refreshment, Hatcher went along the foot of the cliff and found, as one might say now, that he had hit the jackpot. The sediments here were rife not only with single teeth or bone fragments but with many nearly complete skulls, jaws, and even skeletons of animals long extinct. The tremendous Patagonian tides rushed miles up the river, and their erosion of the cliffs was rapidly exposing and disintegrating the fossils buried there some millions of years ago. As Hatcher put it:

> Truly this vast cemetery, which for untold ages had served as nature's burial ground, was now being desecrated by her own hand, with no one present to remonstrate against her wanton destruction of those remains whose very antiquity, it would seem, should have ensured them against such desecration.

Hatcher found the Felton home, unlike that of the governor, to be well up to his standards of comfort, but having come upon the treasures he sought, he wanted to take all possible advantage of them. Having left their outfit with Peterson at Geir Aike, he went back there and with Peterson transported everything to Killik Aike and set up camp near the richest part of the fossil deposit. The two of them worked there from dawn to dusk seven days a week for nearly a month. They then packed up a ton and a half of fossils, added some others collected in a week somewhat farther down the river, and then on 26 June shipped the whole collection on a British schooner with the interesting name of *Bootle*.

By now winter was advancing, and Hatcher determined to examine the coastal region round "Cape Fairweather." (That is another example of Hatcher's anglicizing—this is Cabo de Buen Tiempo on the maps, a curious appellation, as the weather of that region just north of the estuary of the Río Gallegos is rarely fair.) It was the plan of the expedition to collect not only fossils but any other natural history specimens, and at this point in his narrative Hatcher mentions that on this expedition he was primarily the paleontologist and Peterson the zoologist but that as occasion presented they worked together. Fossils were nevertheless the main object, and Hatcher was obviously the head of the two-man party. At first here he did turn his hand, and his guns, to collecting specimens of the present local fauna such as guanacos, which were South American camels related to the llamas and in Hatcher's day abundant as wild animals in Patagonia. He downed two of these, but before he could skin them they were torn up by *caranchos,* or buzzards (although Hatcher wrote of caranchos and buzzards as different scavengers, also differentiated from condors). A considerable collection of mammals and birds was made in this general area. So was a collection of fossils, not of mammals but of marine shells from what Hatcher called the "Cape Fairweather Beds." He believed them to overlie the Santa Cruz beds from which he had collected fossil mammals, although none of these were found in just this region.

On the first of September, as winter was waning, the two explorers moved camp northward to Corriguen Aike, the Indian (Tehuelche) name of a place near the shore and about twelve miles south of Coy Inlet, or Puerto and Bahía Coy. This is the only inlet or port between that of Santa Cruz to the north and Gallegos to the south, but is nearer the latter. (Coy, Coig, or Coyle—different versions of the same name—still appear on modern maps, but I have not found Corriguen or Corriken Aike on any map but Hatcher's own.) Here they found their second and, as regards fossil mammals, last jackpot of the expedition. At the foot of the sea cliff a broad, shelving beach of dark green sandstones exposed at low tide showed embedded and partly protruding the remains of an extraordinary number and variety of fossil mammals, a splendid exhibition of most of the great Santacrucian fauna, now considered early Miocene in age. This extended for about a mile and a half with an average breadth of some three hundred yards. Hatcher wrote jubilantly:

> Throughout eighteen years spent almost constantly in collecting fossil vertebrates, during which time I have visited most of the more important localities of the western hemisphere [an excusable but overdone boast!], I have never seen anything to approach this locality near Corriguen Aike in the wealth of genera, species and individuals.

There were also many footprints on the surface of the rock, including one track extending for about a hundred feet and demonstrating the exact stride of the animal—proof that these animals lived and traveled about in this area before so many of them died and were buried there. It was not possible under the existing conditions to preserve or to collect these footprints, but it is curious that Hatcher, who took many photographs in Patagonia, did not photograph them. Perhaps still stranger is that as far as I know they have not since been sought out and studied. An Argentine paleontologist, Rodolfo Casamiquela, thoroughly acquainted with Patagonia, wrote a whole book on fossil footprints (*Estudios Icnológicos,* 1974) without even mentioning Hatcher's discovery (see Chapter 13).

Collecting here became a sort of daily relay race. As the ebbing tide began to expose the fossil-bearing stratum at the foot of the sea cliff, Hatcher and Peterson would follow the retreat of the water, excavating fossils as they went. As soon as the tide turned, they raced to pick up each specimen in advance of the water and to place it just beyond the level of high tide, still at the base of the great sea cliff. This done each day, as Hatcher wrote:

> We sat on the surface of the talus-covered slope at a safe distance from the waters that dashed furiously beneath . . . while over the sandstones of the beach, from which but a few hours previously we had been excavating the remains of prehistoric animals, there now rolled a sea sufficiently deep for the safe navigation of the largest transatlantic liner. . . . Not many experiences . . . have left themselves so indelibly engraved upon my mind.

They worked there through the months of September and October, a foggy and tempestuous Patagonian spring. At this one locality they collected four tons of fossils, which were hauled to the port of Gallegos and from there shipped to "Sandy Point," that is, Punta Arenas in Chile, on the Strait of Magellan. Hatcher triumphantly remarked of this stratagem, "On account of certain laws prohibiting the exportation of fossils from Argentine territory we were not at all displeased with this opportunity." Hatcher rode to "Sandy Point" on horseback in order to assure the reshipment of the fossils to New York. On they way he had an accident that opened a gash in his head, and he arrived at his destination in bad shape. He had an unhappy time for two weeks in "Sandy Point," but the fossils were cared for and he got back to Gallegos, where Peterson had some more fossils from near Coy Inlet. They then packed up everything, fossils and recent zoological and botanical specimens, started these on their way to New York via "Sandy Point," and on 13 December took off for a trip into the interior of the country, with provisions for eight months.

Along their way to Lake Argentino, the great lake at the eastern font of the Andes, Hatcher's head wound became newly inflamed, and he also developed a high fever. During the succeeding days, he suffered "the most miserable Christmas of my life," but he recovered before long, and they went on to the vicinity of Lake Argentino, through a region that had been previously fairly well explored, by Charles Darwin among others, as Hatcher was aware. In further wanderings the two travelers made extensive collections of recent mammals, birds, and plants, but they found few fossils: some marine shells, one small mammal jaw that Hatcher recognized as Santacrucian in age, and one large dinosaur bone, too cumbersome to be taken along with them. The tour ended by boat down the lower reaches of the Santa Cruz River and then south to Gallegos, where they arrived just five months after their last departure from it. Before definitively leaving Patagonia they took a sight-seeing trip that completely circumnavigated Tierra del Fuego, arriving back at Gallegos on 4 June. They left the next day as a start toward New York, which they reached on 16 July 1896.

Back at Princeton, Hatcher immeditely started preparations for a second expedition to Patagonia. Peterson unpacked and set about preparing their fossil mammals; that is, clearing the specimens from the rocky matrix in which they were still partly embedded. Peterson stayed at that work until the beginning of Hatcher's third and last expedition. As assistant on his second expedition Hatcher took a "kind and obliging young man," A. E. Colburn, who had some experience as a taxidermist but apparently none as a paleontologist. Having had trouble with field transport on the first expedition, this time Hatcher took with him from New York a light "mountain wagon" with double harness. The party of two, plus equipment, left on 7 November 1897, and went nonstop for thirty-two days from New York to Punta Arenas, in Chile. Hatcher bought a pair of work-horses, and the party camped briefly near Punta Arenas, where Hatcher collected some fossil invertebrates and recent plants and Colburn collected some recent birds. From there they went to Gallegos, in order to purchase two more horses and equipment to hitch four horses to the wagon. Hatcher then went to an Indian village to make observations and take photographs, as commissioned by the United States National Museum (now the National Museum of Natural History). He noted that the fraction of whole-breed Indians, even at this relatively early time, was very small. (A generation or so later there were no tribal Indians left in Patagonia.)

Hatcher, accompanied by Colburn, made his way northward to the general area of Lago Pueyrredon, in the rough foothills and eastern slope of the Andes, almost due west of the port of Deseado but south of the headwaters of the Río Deseado. His main purpose was to find fossils of what the Ameghinos called the *Pyrotherium* beds, which Carlos had first located in the valley of the Río Deseado, as noted here in Chapter 4. Hatcher doubted whether this fauna and the strata containing it had been correctly placed in sequence by the Ameghinos, but he found no trace of that fauna either on this excursion or later. He did continue to find some Santacrucian mammals, of which he already had such a rich representation from his first expedition.

Some two months had been spent in this search, a rather futile one as regards fossil mammals although Colburn had made good collections of recent birds and mammals. Now in late April winter was coming on, and Hatcher's first idea of passing the winter in this bitterly cold area seemed less desirable, if feasible at all. However, they had hardly begun their retreat to the coast when Hatcher became desperately ill. He had great rheumatic pain in knees and elbows, all his limbs became swollen, and he developed a high fever. He was completely incapacitated for some weeks, tenderly cared for by Colburn. Finally he could manage with the help of crutches, devised by Colburn, and eventually they reached the port of Santa Cruz. There it was decided that Colburn should wait for a steamer and start back to the United States, while Hatcher, now recovering, would go on horseback and on foot to Gallegos. Despite a Patagonian blizzard, Hatcher did make this apparently impossible trip, and he was in Gallegos when the ship carrying Colburn to Punta Arenas came by. Still Hatcher stayed on in Gallegos, with a persistence which in retrospect seems more nearly maniacal than brave. Then even he gave up and went back to "Sandy Point," where he took a liner for New York, arriving there forty-seven days later on 9 November 1898. The second Princeton Patagonian Expedition cannot really be considered a success, surely not from a paleontological point of view, although Hatcher continued to insist that it was a great success from other points of view.

In somewhat better health, Hatcher was on his way to Patagonia again on 9 December 1898, just one month after he had returned from there. This time he took his brother-in-law O. A. Peterson with him again as his assistant and also Barnum Brown, of the American Museum of Natural History, as in some sense a supernumerary or an undesired co-leader.

Barnum Brown was born in Carbondale, Kansas, on 12 February 1873 (thus sharing a birthday both with Abraham Lincoln and with Charles Darwin although sixty-four years later). In 1897 he joined the department of vertebrate paleontology of the American Museum as an assistant curator, advancing to associate curator in 1911 and to curator in 1927. He retired in 1942 but continued both research and collecting at times into the 1950s. He died on 5 February 1961.

By 1898 Brown was already an experienced fossil collector, working under the direction of Henry Fairfield Osborn. It will be recalled that Osborn had earlier had some idea of sending an American Museum expedition to Patagonia to follow up the lead of the Ameghinos, but that he had been forestalled by Hatcher and possibly also by Scott. Now with the great success of much of Hatcher's work and a splendid collection of Patagonian fossil mammals already at Princeton, Osborn proposed to his old friend Scott that the next Patagonian expedition (it would be Hatcher's third) should be a joint Princeton–American Museum operation, with Hatcher for one institution and Brown for the other. They drew up a written agreement for division of fossils between the two institutions, specifically stating which should have first right to publication on any new genera and species that might be found. Brown readily agreed to sign this, but Hatcher angrily refused. Although Hatcher did accept Brown as a member of his party, it was not a united or happy one.

All three went by ship to "Sandy Point" where Hatcher left the others, but they proceeded by different routes to a rendezvous at Santa Cruz on an agreed date. Brown and Peterson traveled together by wagon, picking up as they went the horses that Hatcher had left along the way from Santa Cruz to Gallegos at the bitter end of his last expedition. After meeting in Santa Cruz the three back-tracked on the route Hatcher had followed so painfully from Lake Pueyrredon the previous winter. Now with two companions sparing him "the irksome duties of a teamster," Hatcher could scour the region for the still elusive *Pyrotherium* beds and fauna found by Carlos Ameghino. This search was again a failure, but some vertebrates (not otherwise specified) and a "considerable collection" of invertebrates of at least two ages were collected. After about two weeks the men separated again, Peterson and Brown with the wagon and outfit returning to Santa Cruz as they had come, and Hatcher, with a saddle horse and pack mule, striking off by himself. After a few days with foul winter weather and some interesting encounters with pumas and deer, fatal to those animals, Hatcher gave up the search and doubled back to Santa Cruz. He passed Peterson and Brown as he reached the trail, and he got to Santa Cruz some two weeks ahead of them. That time was spent in gathering a large collection of fossil invertebrates (almost all mollusks) from the marine Patagonian formation. Soon thereafter he went to Gallegos, and then by ship to Buenos Aires, "leaving Mr. Peterson to continue the work of collecting fossils in the Santa Cruz beds along the coast and to return by the next steamer arriving at Sandy Point and bound for New York."

From Buenos Aires Hatcher took an excursion up the La Plata estuary and Paraguay river to Asunción and nearby villages for the sole purpose of "an absolute rest" before embarking for New York, which he reached on 16 August 1899. Not much later, on 1 September, Peterson also arrived with the last of the Princeton Patagonian collections. Hatcher's summing up was: "We had undergone many hardships and considerable sacrifices in order to accomplish the work. In many respects our success had far surpassed our most sanguine expectations, while we had signally failed in one most important feature of our work, which, however I still hope to accomplish."

The success beyond sanguine expectations was the splendid collection of excellently preserved fossil mammals, all from the Santa Cruz formation, the Santacrucian or (as Hatcher preferred) Santacruzian fauna. The failure was to check the sequence of faunas established by the Ameghinos, and especially the position of their *Pyrotherium* beds and fauna, now called Deseadan. That was Hatcher's main aim on the second and third expeditions, which from that point of view were flatly failures. The blame, if such it should be called, can be placed as much on the Ameghinos' as on Hatcher's side. In Hatcher's day the Ameghinos did not clearly and publicly designate the sites from which their fossils came, although they did so for collectors with whom they were *en rapport*. In this respect it can be said that if Florentino (rarely if ever Carlos) Ameghino was at times unduly secretive, Hatcher was at times both aggressively antagonistic and eagerly self-centered. Hatcher had conversed with Carlos Ameghino (whom he calls "Charles" in notes and correspondence), who would be unlikely to have misled him. (Remember his helpfulness to André Tournouër, as mentioned in Chapter 6.) There may have been lack of understanding—Hatcher did not speak or understand Spanish well, and Carlos Ameghino did not speak or understand English at all.

The clash of ideas and personalities is just an episode that hardly needs more detailed analysis, but one more related occurrence is a matter of interest. In 1903 to 1904 Hatcher and Florentino Ameghino were still quarreling, politely but strongly (Hatcher in English and Ameghino in French) in letters about their indelible differences of opinion. Hatcher cast his last stone. He declared that a single block of rock (matrix) collected by him near the mouth of the Santa Cruz contained fossils that the Ameghinos claimed to be distinctive of three different "horizons," hence faunas, in the sequence. Now that so much subsequent work has shown beyond doubt that the Ameghinos' sequence in this part of the strata was correct (although the ages assigned to them were not), the only likely explanation is that Hatcher or other authorities had misidentified the species in this single block.

Not long thereafter, on 23 February 1903, Hatcher wrote to F. Ameghino suggesting that either he (Florentino Ameghino) "or your brother Charles" (Carlos Ameghino) go with Hatcher to Patagonia "and go over together the complete Mesozoic and Tertiary collections as represented in that region." In a very courteous reply dated "le 14 Janvier 1904" Florentino said that he was too busy to make such an excursion but that if his brother was not away traveling he (Florentino) would ask him (Carlos) to go to Patagonia with Hatcher and if neither of the brothers could go they nevertheless would give Hatcher all the information he needed. It is uncertain whether Hatcher replied to that letter—or even read it—he died a few months after it was written. Thus died also the "hope to accomplish" the "one most important feature" of Hatcher's work in which he had "signally failed."

Here we have gone ahead perhaps too fast, and before summarizing Hatcher's last years we should remember Barnum Brown's having been left in Patagonia when first Hatcher and then Peterson departed to return home. When Hatcher, Peterson, and Brown met in Santa Cruz at what was essentially the end of Hatcher's last Patagonian expedition, Hatcher turned over to Brown "such of my outfit as he desired to use in continuing his work." Although this is not clearly stated in detail, Peterson evidently stayed with Brown just long enough to indicate the areas where Hatcher and Peterson had made such rich finds on their first expedition, and Brown then worked these rich fossil beds alone on behalf of the American Museum.

Brown by himself, as he preferred to be in the field, made a good collection of Santacrucian fossil mammals for the American Museum. Apparently derived entirely from localities which Carlos Ameghino, Hatcher, or both had already thoroughly prospected, it is not surprising that only one species in Brown's collection was definitely new. In accordance with the agreement which Brown had signed but Hatcher had not, the well-preserved skull of this species was described by Brown in the 8 July 1903 issue of the *Bulletin of the American Museum of Natural History*. It was a small glyptodont, one of the relatives of the armadillos encased in bony armor. The genus *Eucinepeltus* had been named in 1891 by Florentino Ameghino on the basis of a specimen found by Carlos, but Brown's *Eucinepeltus complicatus* has been generally accepted as a species distinct from that of the Ameghinos.

Brown is most remembered as a collector and describer of dinosaurs, although he also made important collections of fossil mammals, especially in Greece and in what was then India (but in an area now divided between India and Pakistan). He never returned to South America, but in the 1950s, after his retirement, he spent about a year in the Peten region of Guatemala. There, on his own but with the backing of his second wife Lilian (née MacLaughlin), whom he had married in India, and some local assistance he made a collection which included a mixture of mammals from North and South America. These mingled in that region during the Great Interchange of faunas in the last two million years or so.

Upon returning to Princeton in 1899 Hatcher's first consideration was that publication of the results of his work in Patagonia should not be scattered through journal articles and other miscellany but gathered into a large set of volumes uniform in format but by several specialists in the various sections. Scott agreed, but this undertaking would obviously be very expensive and beyond the means of the department or university. An appeal was made to J. Pierpont Morgan, who (to the delight and, one feels, the covert surprise of all concerned) agreed to endow a publication fund adequate for the purpose. (The preparation and publication of the "Reports of the Princeton University Expeditions to Patagonia, 1896–1899" will be covered in Chapter 8, since responsibility for seeing the project through devolved upon Scott.)

Nevertheless all was not well. As had happened in his relations with O. C. Marsh at Yale, Hatcher felt, with considerable reason, that his accomplishments were not fully appreciated and that he had additional abilities that were underrated and given too little opportunity. The result now was a letter to Professor Scott, written on 17 November 1899:

> After our conversation today and in view of the fact that I am about to make application for a position elsewhere (though where I do not know) I consider it my duty to tender to you my written resignation as your assistant. I would have taken this step before except for a consideration of the possible injurious effect it might have had in securing the money necessary for the publication in a proper manner

> of the results of the Patagonian work. For as you know even a suggestion of friction is frequently quite sufficient to influence negatively those who would otherwise have given freely. I confess also to having been partially deterred by a hope that some joint action by yourself and Prof. Osborn might be taken to explain matters and make the action unnecessary. Now that the money is already pledged I no longer feel any restraint on that score and since in the latter more personal matter my hopes have not been realized, there is nothing left to me except to either submit to treatment which I believe I have not merited or resign. I therefore enclose my resignation to you. . . .

Hatcher did soon find a position as curator of vertebrate paleontology at the Carnegie Museum of Natural History in Pittsburgh, of which Dr. W. J. Holland was then director. There Hatcher had an able and congenial group to work with, including among others his brother-in-law Peterson and Charles Gilmore, who was destined to become curator of fossil reptiles in the United States National Museum. Yet once again all was not harmony, as is obvious in the following extract of Hatcher's report for November 1900, as curator to the director.

> I have been deeply grieved to find that my conduct of the Department placed in my charge has been so unsatisfactory as to call forth the personal abuse visited upon me, by yourself on Nov. 7th. I am also much affected by the further abuse you saw fit to administer on Nov. 8th when you called me a jack-ass and a d—d fool. Such language, it seems to me, cannot but tend to destroy that harmonious enthusiasm & interest so essential to the welfare of this institution & should therefore be discouraged.

Thus Hatcher's pioneer personality boiled over again, and one thinks that is his way of saying, "When you call me that, smile!"

Ailing much of the time, but furiously and persistently busy (collecting and preparing dinosaurs for the most part), Hatcher did spend his pitiably few remaining years at that job for the Carnegie Museum. On 3 July 1904 at the age of forty-two he died of typhoid fever. When the secretary of the Carnegie Institute sent news of his death to Andrew Carnegie at Skibo Castle in Scotland, Carnegie expressed "keen sorrow" and added, "If Hatcher's wife and family are not provided for, we must surely do something for them. Please look into this."

Hatcher fell out, at one time or another, with the great men of his profession, notably Marsh, Osborn, Scott, and Holland in succession. Marsh predeceased him, but Hatcher dedicated his major work, the Patagonia narrative, "To the Memory of Othniel Charles Marsh." After Hatcher's death Osborn wrote of "the rare and noble spirit of John Bell Hatcher."

Scott wrote that his death was "an irreparable loss' to American paleontology. Holland found that "he was a most charming companion, and when he could be prevailed upon to unbend and relate the story of his adventures in strange and distant places, the listener found his companionship fascinating."

Much that was interesting in his life has been left out here or only mentioned in passing, as not directly relevant to the theme of the book, but two other points may be briefly noted. Next to the Patagonia narrative the main publication credited to Hatcher is *The Ceratopsia* (horned dinosaurs), published in 1907 by the United States Geological Survey as "by John B. Hatcher, based on preliminary studies by Othniel C. Marsh, edited and completed by Richard S. Lull." Thus this work went from hand to hand as forced by death. Marsh started the study, based on collections made by Hatcher. When Marsh died in 1899, Hatcher himself took up the project but was still rather far from completing it when he in turn died in 1904, and then on the advice of Osborn it was turned over to Lull, who completed it and was the author of much of it but made Hatcher the senior author.

The last point to be made in this chapter is that Hatcher believed South America had at some time been attached to Antarctica, and through Antarctica to Australia. Hatcher made great efforts to be included in one of the two British Antarctic expeditions, one English and one Scottish, being planned in 1901 to 1903, but he was unsuccessful in this. His idea seemed fanciful at the time, and was then really no more than a weak hypothesis. Yet it turns out now to have been probably correct.

Notes

Epitaphs or memoirs of Hatcher by Scott and by Holland summarize and eulogize his life as a whole. The selectively abbreviated account of his three Patagonian expeditions in the present chapter is of course based almost entirely on his long, detailed account in the first volume of the "Reports of the Princeton University Expeditions to Patagonia, 1896–1899." Some interesting and important aspects of his life and character have been derived from parts of his correspondence not published in the basic references and kindly made available to me from the departments where he worked in Princeton University and the Carnegie Museum.

Scott

The preceding chapter made it evident that W. B. Scott was involved with South American paleontology from the start of Hatcher's expeditions there. The present chapter will place that involvement within the general framework of Scott's long and varied life and will carry the story forward beyond Hatcher's time.

Despite his wide travels and close associations with other scientists on three continents, William Berryman Scott became so identified with Princeton, the town and the university, that it seems anomalous that he was not born there. He was in fact born in Cincinnati, on 12 February 1858. (Another 12 February boy, like Darwin, Lincoln, and Hatcher!) His parents, William McKendree Scott and Mary Elizabeth Scott (née Hodge) had met in Princeton while W. M. Scott was studying at the Princeton Theological Seminary, and after their marriage had moved to Danville, Kentucky, where Scott *père* became pastor of the Presbyterian Church and a professor at Centre College. (It was there that the son who became General Hugh Lenox Scott was born—the one considered by most historians, mistakenly in my opinion, to be the only famous member of this family. He fought in the Indian wars, then was successively adjutant general of Cuba, governor of the Sulu Archipelago, superintendent of West Point, chief of staff of the U.S. Army, held a command in France during World War I, and finally retired in 1919, but did not stop his varied activities until he died in 1934 at the age of eighty-one.)

From Danville the Scott family moved to Cincinnati and thence briefly to Chicago, but around 1861 the failing health of the father prompted them to move back to Princeton, where Mrs. Scott's family was still living. It was there and as a member of that family that William Berryman Scott grew up, and much as he traveled elsewhere Princeton remained usually his physical and always his emotional home for the rest of his long life. It was a matter of justifiable pride but modest expression that he was a direct descendant of Benjamin Franklin through his mother and also through the Bache family, eminent in Philadelphia. He was also a descendant of Catherine Wistar, the sister of Caspar Wistar who was, as Scott later expressed it, "in point of time, the first of American vertebrate paleontologists." (The basis for this has been mentioned in Chapter 1, above.)

After somewhat ragged elementary schooling, some at home, some in Philadelphia, and some in Princeton, in 1873 William Berryman Scott entered the College of New Jersey, the forerunner of Princeton University. Scott was not particularly popular among the students, but he early made one friend, Frank Speir, who had the student nickname of "Sally" and with whom for more than fifty years Scott had "a warm and unbroken friendship." Later, in his junior year, he formed a similarly close friendship with Henry Fairfield Osborn, first nicknamed as "Harry" and then as "Polly." (In later years even those most familiar with that great man would not have dared to address him less formally than as "Professor," which continued even after he was in fact no longer a professor.) Scott's own nickname from childhood continued to be "Wick" to close friends of his own, or an older, generation.

An anecdote about the three college friends, Sally, Polly, and Wick, concerns a casual incident that eventually had a tremendous influence on the history of science, or more specifically of paleontology. Although it happened well over a century ago, it is still being repeated and is remembered by at least the older or more historically minded paleontologists of today. Scott related it in the first sentences of his *History of Land Mammals in the Western Hemisphere,* published in 1913:

> One afternoon in June, 1876, three Princeton undergraduates were lying under the trees on the canal bank, making a languid pretence of preparing for an examination. Suddenly, one of the trio remarked: "I have been reading an old magazine article which describes a fossil-collecting expedition in the West; why can't we get up something of the kind?" The others replied, as with one voice, "We can; let's do it.

After great effort and persuasion the expedition did indeed occur in 1877, although not just in the form anticipated by the three idlers on the canal bank. As authorized and financed by the trustees of the college, it eventually got under way with two leaders: Dr. C. P. Brackett, professor of physics, who directed the scientific work, making collections in botany, zoology, paleontology, and mineralogy, and a General J. Kargé, who was in military command, assigned to train the party for combat and lead the fighting with Indians—fighting which did not occur. There were sixteen students selected by competitive examination, two photographers, and two cook-teamsters. Oddly enough in this egregious medley the three Princeton undergraduates who had dreamed the whole thing up only for the fun of going out and picking up some fossils really did just that. Scott, or Wick as he should be called in this company, found the first fossil, a leg and foot of a tiny ancestral horse.

That expedition was followed by others, after graduation from Princeton, which need not be enumerated here. The important point was that Scott, who had first considered studying for the ministry and then medicine, and Osborn, who had expected just to take over management of his wealthy family's finances, both dropped those plans and determined to become naturalists, preferably vertebrate paleontologists. The third daydreamer, Speir, continued close friendship and did visit some later expeditions, but only when he could take time off from his law practice.

The three friends graduated together with the Princeton class of 1877. By then Scott and Osborn were both eager to continue their studies in geology and biology (the overlap of which is paleontology), but the route for them was not yet clear. First of all, at that time the centers for advanced study in academic sciences were European, especially German or English or both. Scott therefore took off for England, with Henry Fairfield Osborn's father, William Henry Osborn, a railroad millionaire, paying the expenses. Scott was vague as to just what course to pursue and where to do so. Writing later, he said that in effect all he was told was "Go to England and study something with somebody, and then go on to Germany and study something else with some other body."

Only twenty years old and in a culture alien to him, Scott was in a daze, but he had the good fortune to have a letter of introduction to Thomas Henry Huxley. Huxley firmly told him, "What you ought and need to do, is to come here to South Kensington and take my course in the Royal College of Science." Scott followed that advice, and it set him firmly on the track that he would follow all the rest of his life. On Huxley's advice Scott also went to Cambridge, where he worked informally with Francis Maitland Balfour, a brilliant student and professor of anatomy and embryology. It was here that Osborn joined Scott, Osborn having been reluctantly allowed to leave the counting house and plunge back into academia. Osborn went to London, where he in turn took Huxley's course and met other great scientists of that time, briefly including Charles Darwin. One of the many anecdotes about Osborn's well-deserved self-esteem after he had become famous in his own right is that when he first met one of the young acolytes he would say, "Don't be nervous. I know you feel just as I did when I was young and shook hands with Darwin!"

After finishing Huxley's course, Scott took off for Germany with his mother and studied mainly at Heidelberg for another year. He then went back to Princeton, began his research and teaching career there, completed his dissertation, and received his Ph.D. in 1880. Thereafter in spite of numerous travels he remained on the Princeton faculty, establishing and heading a department of geology, until his nominal retirement in 1930, "nominal" even at an age over seventy as he continued to work until his death not far from twenty years later.

When Scott was twelve years old in 1870 he had met a ten-year-old girl named Alice Post, a distant relative, being "a niece of my [Scott's] Uncle Wistar's wife." According to his later story he then and there determined to marry her when they were both a bit older, and on 15 December 1883 the marriage did indeed take place, lasting happily for many years until in old age death intervened. Scott's appreciation (as well as his strong sense of decorum) appears in his final mention of her in his long autobiography:

> One does not like to speak publicly of his family life, but I must give testimony that no man ever had a nobler helpmeet. Had she been a selfish or small-minded woman, she might have sadly hampered or even put a stop to my work, but she has ever been a tower of strength and the best of comrades. Of what she has been to her children and of their devotion to her, I must let them speak.

This subdued tribute resembles that of Charles Darwin to his wife Emma but is even more reticent.

Installed at Princeton, Scott continued the collection, study, and publication of fossil mammals of the Cenozoic era (the "Age of Mammals") of western United States. Even after old age curtailed his personal collection of them he went on usefully and extensively as an accumulator, compiler, and monographer in that field. The continuity in his professional life was nevertheless abruptly broken into near the end of the nineteenth century when Hatcher suddenly, so to speak, dropped many hundreds of Patagonian fossils in his lap. Having helped briefly with first steps toward the scientific utilization of this windfall, Hatcher took off for Pittsburgh, dropped the subject of South American mammals, and went on brilliantly in quite a different career in vertebrate paleontology, cut short all too soon by his early death. Arrangements for publication on the fossil mammals from the Patagonian expeditions necessarily remained Scott's responsibility. An account of the expeditions themselves could be written only by Hatcher, and he did write that account in detail, as summarized here in the preceding chapter. With hardly any exceptions, all the fossil mammals collected, although from a number of different regions and localities, were from a single major fauna, that of the Santa Cruz formation and of an age called Santacruzéen by Florentino Ameghino, now generally labeled technically as the Santacrucian land mammal age. (As will appear later, Ameghino's placing of his Santacruzéen in the geological time scale is not now accepted for the Santacrucian.)

The collections of plants and animals were outside of Scott's sphere, and also that of the present book. The collections of fossil marine invertebrates, mostly mollusks, were not one of Scott's specialities, but they did belong in his department. They were also crucial for the placing of the Santacrucian fauna in the geological time scale. This part of the collections was assigned for study to Arnold Edward Ortmann, who was then curator of invertebrate paleontology at Princeton. He was born in Magdeberg, Prussia, on 8 April 1863 and had studied at Jena and Kiel in Germany, then at Strasbourg (then also German but earlier and later French), and finally received his Ph.D. from Jena in 1885. In 1894 he moved to the United States and to Princeton University. He followed Hatcher to the Carnegie Museum in 1903 and in 1909 joined the faculty of the University of Pittsburgh, where he continued until his death on 3 January 1927. His volume on the Tertiary invertebrates of the Princeton Expeditions to Patagonia was completed and published at Princeton in 1902, while he was still there.

The plan for uniform publication of the "Reports of the Princeton University Expeditions to Patagonia, 1896–1899," like the plan for the expeditions themselves, seems to have been Hatcher's idea, although this is not entirely clear in the available documentation. It is clear that he expected, or was expected, to study and write the Reports on two of the major groups (orders, in classification) of Santacrucian mammals: the Marsupialia, related to some degree with the major mammalian elements in the Australian fauna, and the Litopterna, a large group of hoofed mammals (ungulates) that originated in South America, became extinct there late in the Cenozoic, and are not known from any other continent. It is not recorded whether Hatcher did begin study of either or both of those groups, but in any event he did not proceed far enough to produce anything publishable about them. Scott therefore took on the Litopterna as one of the numerous groups that he would study himself and assigned the Marsupialia and also a group of specialized South American ungulates, the Typotheria, to his colleague at Princeton, Sinclair.

William John Sinclair, son of Samuel Fleming Sinclair and Ellen (née Milliken) Sinclair, was born in San Francisco in 1877. He studied in California schools and attained his Ph.D. at the University of California, Berkeley, in 1904. In that same year he went as a fellow to Princeton, where he successively became an instructor in geology in 1905, assistant professor in 1916, associate professor in 1923, and professor in 1930 as a successor to Scott in the sense that Scott officially retired that year, although Scott continued working in the department and outlived Sinclair by twelve years. Sinclair married Della A. Coleman in 1914. They had no children, but Sinclair was paternal, almost *in loco parentis,* to his student Glenn Lowell Jepsen, always known as "Jep." Jep joined the Princeton faculty in 1930 and became the Sinclair Professor of Vertebrate Paleontology in 1946. He was born in Lead, South Dakota, on 4 March 1903 and died in Princeton on 15 October 1974.

Early in his career Sinclair worked in the field and also did office research in connection with the American Museum of Natural History programs, especially with Walter Granger on early North American fossil mammals in 1913–1914. He never visited South America or studied any South American fossils except the Marsupialia and Typotheria, as noted above. Having done with those, he worked on North American fossil mammals, especially those of the Oligocene in the Big Badlands of South Dakota. In the study of early fossil mammals in western United States, Scott-Sinclair-Jepsen formed a closely connected sequence, but as regards South America the sequence tapered off with Sinclair and ended without Jepsen, who did no work on South American fossil mammals, although many were in his care, mostly from Hatcher's collections and a few subsequently.

Now I have introduced and rather hastily ushered out two investigators who are well remembered and highly honored in other fields but are only ancillary to the present subject. Let us return to the problems faced by Scott for dealing with the great new collection of South American fossil mammals that was in his hands. These specimens had to be individually identified, then grouped or classified in a systematic way according to their most likely relationships to each other and to other groups of animals, and then fitted somehow into a geological and geographic system. The results of this work had to be reduced to writing, illustrations had to be made, and the whole to be published as fully and usefully as possible. Those procedures, each more complex than may appear in such statements of them, fall necessarily and most simply into two groups or processes: first study the specimens, then publish the results.

Scott went about these two overall aims in what may seem an odd sequence, but one that worked out well. Most of the taxa (families, genera, and species) in this extensive fauna had been first recognized and named on the basis of type specimens then in Argentine museums and especially in the extensive still privately owned collection of the Ameghino brothers. Many or indeed most of them had been poorly illustrated or not figured at all in publictions. Therefore to study the new—and on the whole better and more complete—specimens required first-hand knowledge of those types and good illustrations of them. That meant a trip to Argentina, open access to the specimens, and skillful photographing of them. For publication, photographs and drawings of the Hatcher collection must also be made. That could be done in Princeton, but to publish them well required at that time expert lithography of the drawings. The text must be handset in type, and each division of the large series must be printed and put together by an expert and responsible publishing firm. Scott knew that the best lithography and publication could be obtained at a fair price in Germany, where he was well acquainted. Therefore a trip to Germany was necessary, and he went to Europe first as a start of this monumental project.

In Paris, Scott attended the Universal Exposition of 1900, there met his friends Professor and Mrs. Osborn, and also became acquainted with Karl Alfred von Zittel, a professor at the University of Munich and one of the most famous paleontologists of his times. In Frankfurt he engaged the lithographic firm of Werner and Winter to make the plates for the Patagonian reports. Over the years, which extended from 1903 to 1932, these were beautifully done from drawings for the most part by Bruce Horsfall and F. von Iterson. (Some photographic plates were made in the United States, as were the text figures.)

For printing and publication, von Zittel had recommended E. Nägele in Stuttgart, where Nägele had taken over an old and esteemed publishing house with the elegant Teutonic name "Schweizerbart'sche Verlagsbuchhandlung." Nägele agreed to handle publication of the reports except for the lithographic work already assigned to Werner and Winter. Thus each part of the long series of publictions came to have on its cover (when unbound) this array:

J. PIERPONT MORGAN PUBLICATION FUND
Reports of
Princeton University Expeditions
to Patagonia, 1896–1899
J. B. HATCHER IN CHARGE
EDITED BY
WILLIAM B. SCOTT
BLAIR PROFESSOR OF GEOLOGY AND PALÆONTOLOGY,
PRINCETON UNIVERSITY
[then the volume and part numbers, their subjects,
and the name of the author of this part, either
Scott or Sinclair for the parts on fossil mammals,
followed by, at the bottom,]
PRINCETON, N.J.
THE UNIVERSITY
STUTTGART
E. SCHWEIZERBART'SCHE VERLAGSHANDLUNG
(E. NÄGELE)
[Year of publication]

(The attentive reader will note that the usual but redundant "buch" has been excised from "Verlagsbuchhandlung," which simply means "book publisher business.")

His arrangements made in Germany, Scott went to England and attended the meeting of the British Association for the Advancement of Science (affectionately known as "the British Ass"). Although now finally on his way to Argentina, he had arranged for a considerable stay in London, especially to complete preparations for photographing the crucial type specimens of Santacrucian mammals in La Plata. He had purchased a camera in the United States for the necessary technical photography, but he did not have all the needed paraphernalia and supplies. He also was quite innocent of how to do such scientific photography, and so were most professional photographers. In London he bought a supply of glass plates (there was no suitable film then) along with "developers, dishes, red lanterns, etc." He also employed a professional to instruct him in the use of all this. As he observed in his "Notes for an Autobiography," first written for his children but later somewhat abbreviated and modified for publiction as a book: "I got the lessons all right, but my teacher was very much bored with me, for I had to come to him after business hours, when he wanted to knock off work; but I learned enough to accomplish an immense amount of work in Argentina."

While still in London he also accomplished an immense amount of visiting of old friends, meeting new ones, going to sessions of scientific societies, and the like. One of the new friends was F. P. Moreno, then director of the La Plata museum, who was very cordial to him and gave him letters to the officials of the museum. One can be sure that Moreno gave him no letter to Florentino Ameghino, for the bitter enmity between Moreno and Ameghino never abated. Fortunately Moreno stayed on in London and did not return to Argentina until Scott had left there. Otherwise, the situation would have made it impossible for Scott, as it earlier had for Richard Lydekker, to have the access he needed to the collections both in the La Plata museum and at that time still in Ameghino's hands.

At this point in the autobiography written for his children and later modified for publication Scott recounted an anecdote involving Moreno and Hermann Burmeister (whom he called a "disreputable old German"). As an example of "Burmeister's tricks" Scott noted that "Moreno was a good deal of an amateur" but that on a collecting trip to Patagonia he found a particular prize: "the partially preserved skull of a new and very peculiar animal from the Santa Cruz formation, which he intended to name and described himself." However, "Old Burmeister" managed to get a good look at the specimen while Moreno was away, and he published a description of it and named it *Astrapotherium magnum*. Moreno was naturally furious, and he later proposed a different name: *Mesembriotherium*. However, under the code of zoological nomenclature the first publisher of a name wins, honestly or (as in this case) not, and the name proposed by Moreno is an invalid junior synonym, forgotten except for anecdotes like Scott's.

After his stay in England Scott sailed for Argentina, arriving some weeks later at Ensenada, the port of La Plata, as most ships then did, and continuing to Buenos Aires from there by train. There was great newspaper publicity about the arrival of such a famous scientist from North America, and when he objected he was assured that it would benefit the La Plata museum, which was dependent on government funds. After a few days he went down to La Plata, because all the specimens he wanted to study and photograph were there. He was unfavorably impressed with the city, which was still under construction as the capital of the large province of Buenos Aires. (The city of Buenos Aires itself is the capital of the country as a whole, and as a federal district it is distinct from the province and its capital, much as Washington and the District of Columbia in the United States are distinct from the states, each with its own state capital.) The La Plata museum, however, was essentially completed, a splendid structure that still stands with only minor changes since Scott was there.

Thanks to Moreno's letters of introduction, Scott was given a bedroom, bathroom, and study in the living quarters in the museum building. (These were still in use for guests at least as late as the 1930s but have since then been adapted for staff offices and exhibition.) Scott was thus a guest of the museum, and in the absence of Moreno, a Sr. Catani, as acting director, was hospitable and helpful. At first Catani and Scott tried to communicate in French, which neither of them spoke well, but it turned out that despite his Italianate name Catani was Swiss and that German, in which Scott was also fluent, was his native language. In fact virtually the whole scientific staff was either German or Swiss German (*Schweizerdeutsch*), and Scott, with his doctorate from Heidelberg, got along famously with them.

An exception to this good will was Alcides Mercerat, a Swiss surveyor whom Moreno had brought in as paleontologist when he exiled Ameghino from the museum. Scott came to distrust Mercerat as the result of a curious story, told in his "Notes for an Autobiography" but not mentioned in the later published volume. While he was in La Plata, he was offered for sale some Santa Cruz fossils in Mercerat's possession. Scott examined them, but found them so inferior to specimens collected by Hatcher that he did not care to buy them or even to accept them as a gift. Later, back in Princeton, shortly after Hatcher had died, Scott received from Mercerat "a threatening, blackmailing letter, in which he laid claim to the Santa Cruz collections which Hatcher had brought home." Scott went on to write that in this letter Mercerat "said that he had collected all those fossils himself, had packed them up and stored them in a warehouse in Gallegos. Hatcher had learned of this collection, had bribed the custodians to let him take it and had shipped it to New York." Scott discussed this with O. A. Peterson and with President Wilson (then president of Princeton, not yet of the United States), who said, "There isn't a court in Christendom that would consider Mercerat's claim." Scott wrote to Mercerat merely saying that he (Mercerat) had been "misinformed" and that all the Princeton collection had been made by Hatcher and Peterson. That ended the matter.

But to go back to Scott's stay in La Plata, as soon as he was settled in the museum, R. Lehmann-Nietsche, the anthropologist there, took him to Ameghino's house and introduced him. This was during the period when (as here discussed in Chapters 4 and 5) the Ameghinos had been forbidden by Moreno even to enter the museum and Florentino had to operate a librería, selling slate pencils to children, as Scott put it, in order to keep Carlos in the field and to give Florentino time between sales to write his increasingly voluminous works. Scott quite properly considered it shameful that so eminent a man should be reduced to this degrading situation, and he expressed the greatest admiration for Florentino, forging ahead as best he could against such difficulties. Scott wrote later, "I do not know a finer example of courage and devotion under the most adverse circumstances."

About his first meeting with Florentino Ameghino (he never did meet Carlos), Scott wrote:

> I was received with the utmost cordiality and all the resources of the collections were placed unreservedly at my disposal. I feel assured that Ameghino's liberal and generous treatment of me was due to kindness, but he also found an advantage to himself, as I learned a few weeks later. He told me that he was glad that I had come to study his collections and report upon his work, because "now the paleontologists of Europe and of America will recognize and admit that I have done my work loyally." Probably, however, that was an afterthought. He received me with great dignity and made no apology or explanation of his occupation and mode of living; yet he evidently felt the incongruity of it all.

Scott's own insight and generosity of spirit did lead him to spread Florentino's reputation in North America and Europe as a great scientist and noble man, yet there was an inevitable irony. Although he never diminished his praise and admiration for Ameghino, Scott refuted and one might say annihilated precisely the three major, general points that Florentino Ameghino most emphasized and most expected to give him a place among the immortals of science. These were mentioned in Chapter 5. Later in this chapter will appear how and when Scott gave them the *coup de grâce*.

At La Plata Scott's heavy but congenial labors were soon established as a steady and productive program. Fox six days a week (Sunday excluded) he spent mornings working at the museum. Then he had luncheon at the Hotel Mainini with Lehmann-Nietsche, after which he walked to the building that was Ameghino's home, store, museum, and work shop. Both there and at the museum Scott's main tasks was to annotate, measure, and photograph the most important specimens, which generally were the types—that is, the specimens on which the published names of genera and species had first been based. For some time Scott "merely made the exposures and developed the plates," as he wrote. No slight job in itself, but for a time the prints from the plates were made by a Sr. Bruch (apparently a technician at the museum). Shortly, however, Bruch was not available and Scott had to do his "own printing, which was then a much more elaborate and tedious process." He used the "already then old-fashioned 'printing out paper,' in which the picture was produced by exposing the paper under the negative in strong and direct sunlight. When the image was fully printed, it was necessary to fix the print by dissolving out the excess silver and then tone it in a gold solution, which removed the objectionable red colour of the prints." No wonder that this "added very much to [his] labours."

In the evenings he studied Spanish and did some translations of Spanish technical works in paleontology and geology, of which he made a large collection. As all scientists know, it is far easier to read technical literature in a foreign language than to read other sorts of literature or to converse. Scott and Ameghino conversed only in "bad French." Ameghino could read but not speak English and German but was fluent in French of a sort. He had, after all, lived for some time in France and had a French wife. Scott, whose conversational French was exiguous, found Ameghino's fluent but phonetically peculiar. According to Scott, Ameghino pronounced the numerous e's in French, sounding them as *o*'s, for example saying "autro" for "autre." This seems unlikely and perhaps Scott's ear failed him—also one notes that if a Frenchman pronounced "autro" it would become "otro," which is the equivalent word in good Spanish and not just bad French. These trivialities come up in Scott's account in introduction to the following statement:

> Gradually, however, we reached a basis of mutual understanding. . . . Every afternoon at half-past four or five, Ameghino would appear with a tray, bearing tea and the necessary paraphernalia for preparing and consuming it. I could not bear to dull his enthusiasm by telling him how much I disliked that innocent beverage and I therefore solemnly imbibed one or two large cups. During our tea-drinking, we discussed the problems of paleontology and squabbled with the most perfect amiability, for we hardly ever agreed about anything and yet we never lost our tempers and always kept the discussion on the purely objective plane.

It would be folly to doubt either Scott's word or his memory, but there are two oddities in that statement that are hard to take quite literally. First, is it not possible that the "tea" disliked by Scott was not tea in the English sense but the so-called "Paraguayan tea" or in Argentine Spanish *mate,* an entirely different beverage but much the more usual one in Argentina? (*Maté* was the only hot beverage that Carlos Ameghino ever drank, but we do not have this information about Florentino.) And second, is this not the only recorded example of anyone ever disagreeing with Florentino Ameghino and keeping discussion "on the purely objective plane"?

Whatever their differences of opinion were, Scott stated that he came to La Plata with prejudice but "soon came to respect his [Florentino Ameghino's] work, as I had not done before. Nearly all of the known Santa Cruz fauna had been named and classified by him and our great collections added but few new forms that had not been described before. The immense value of the work done by Hatcher and Peterson consisted in their skill as collectors, in finding and extracting skeletons of many creatures which had been known before only from fragments, or scattered bones." That is true, but it does less than justice to Carlos Ameghino, who is here not even mentioned although the inferior quality of his collecting is involved obliquely, as it was directly by Hatcher. It was nevertheless Carlos who discovered almost all of the known genera and species not only from the Santa Cruz formation but also from other formations that Hatcher and Peterson failed to find. Hatcher and Peterson were more experienced and skilled collectors, and they had better equipment, but Carlos Ameghino was the more successful pioneer and the more *criollo* (native, adapted to the country), as an Argentine would put it. It was the primacy of the Ameghinos' work that required Scott's long labors with the materials in La Plata before going back to Princeton to work with Sinclair on the revision of this classic fauna.

Scott made several trips to Buenos Aires for rest and recreation, also for a quick look at some of the Pampean fossils, then poorly housed there. Despite these recreations, the stay in Argentina was a trying one. As the work drew to a close Scott made duplicate prints of all the photographs of types in the Ameghinos' collection and put the duplicates in an album which he presented to Florentino. On 25 October 1901 he sailed from Ensenada for New York and eventually Princeton.

One aftermath of Scott's stay in La Plata merits mention here, and in slightly varying formats it was thereafter a lifelong standard in his frequent conversational anecdotes. He was given a gold medal by a geographical society, and among the encomiums one linked it to his "extensive travels in Patagonia." His anecdotal version was that he was the only person who even received a medal for exploration without the usual prerequisite of visiting the region explored. (He never did go to Patagonia.)

In 1903 the long series of reports on the Princeton Expeditions to Patagonia began to pour off the press with Hatcher's narrative as the first volume. These publications included a major contribution to the natural history of Argentina, but their main thrust was a complete revision of the Santacrucian fossil faunas. This not only revised the fauna from a taxonomic or classificatory point of view but also included the nearly or quite complete skeletons of a number of mammals hitherto known only from fragments. Scott always maintained his great admiration for Florentino Ameghino, which he warmly expressed in his autobiography, published in 1939, twenty-nine years after Florentino's death. Nothing could contradict or undermine the tremendous accomplishment of the Ameghino brothers under such difficult circumstances, but, as has already been mentioned, Scott and Ameghino did differ in some respects in their friendly afternoon chats over tea (or was it *maté*?). Some mention of these was made in Chapter 5, but the major differences piled up later as the Princeton Reports were issued. I have mentioned that they eventually annihilated what had seemed to Ameghino the most important generalizations and deductions of his own work. At this point we may consider more specifically just what those Ameghinian points were, and how the reports disproved them.

The most excusable or in theological terms the merely venial fault in Ameghino's work, one into which any really active classifer and namer of many fossils has been likely to fall to some degree was his failure to understand that species of fossils, as well as of living organisms, are populations, not individuals, and that they are sure to include individuals variable in many of their characteristics. If that were not true, evolution would not occur in the ways that it usually or probably always does happen. The most striking example in the Princeton Reports, and the one that Scott especially mentioned in the autobiography, is in Volume VI, Part II, "Toxodonta of the Santa Cruz Beds," by Scott, published in 1912. As mentioned in Chapter 1 of the present book, the genus *Toxodon* was named by Owen on the basis of specimens collected by Darwin. Its age was relatively late, as it became extinct only a few thousand years ago, but further discoveries showed that it had many close relations millions of years older, forming a family Toxodontidae as part of a still broader, more varied group, the suborder Toxodonta.

Scott's research on the Toxodontidae in the Santa Cruz collections found that the most common genus of the family in those beds was *Nesodon*, named by Owen in 1846 in the same publication as *Toxodon*. When Scott reviewed all the specimens available in the early 1900s he found that eleven different names had been applied to this single genus up to 1891. Of these, three used by Moreno and Burmeister were valid genera but not properly distinguished from the earlier named *Nesodon*. However, six by Ameghino and one by Mercerat were failures of interpretation of generic characters or were invalid synonyms of *Nesodon*. Furthermore, as Scott later stressed in his autobiography, twenty-eight different names had been used for what Scott's researchers showed to be a single species of *Nesodon*. Of these, sixteen were first published in 1891, all in the same paper by Alcides Mercerat. Obviously that only confirms the general opinion, then and ever since, that Mercerat was not a competent paleontologist. However, eight were named by Ameghino, or transferred to this genus by him, and this cannot be ascribed to inexperience or incompetence but only to a mistaken view as to the latitude of variation within a species. The name considered valid under the code of nomenclature on the basis of priority is *Nesodon imbricatus* Owen, 1846, but it is noteworthy that Owen simultaneously published another name, *Nesodon Sulivani,* which is a synonym of *Nesodon imbricatus.* Moreno and Burmeister also contributed two invalid names to this mess.

A second and much more serious generalization and systematization of his studies by Ameghino, also mentioned in Chapter 5, was also definitely and finally contradicted largely by results of the Princeton Expeditions to Patagonia. It has been noted that Hatcher collected marine invertebrates from the Patagonian formation and that this immediately underlies and even at the top somewhat intergrades with the base of the Santa Cruz beds. Ortmann's study of those marine fossils in the Princeton Reports indicates that the age of these marine fossils is probably early Miocene or possibly somewhat but not much older. The Princeton expedition found no mammals below the Patagonia formation, but Carlos Ameghino had found a whole series of them, one fauna, in what are called the *Colpodon* beds by the Ameghinos but are now called Colhuehuapian, which immediately underlies and marginally intergrades with the bottom of the Patagonian formation. This sequence was supported also by Tournouër, and he further sent marine Patagonia formation fossils to France where they confirmed the sequence and ages involved. Thus the evidence of the Princeton expeditions, combined with that of studies in Paris and ironically with the field evidence of Carlos Ameghino, put it beyond later serious question that the Santa Cruz beds must be no older than early Miocene, as against the late Eocene age given them by Florentino Ameghino, and the *Colpodon* or Colhuehuapian must be later than the early Eocene age ascribed by Florentino. In his final summing up of the Princeton Reports Scott paid little attention to the pre-Patagonian, as he had in hand no mammals earlier than the Santa Cruz, but it followed from the new orientation of the Patagonian sequence in geological time that all of the ages Florentino assigned to faunas were likely to be much too old, as later and more modern methods of timing have indeed proven.

The correction of the third great generalization in Florentino Ameghino's work cannot be so distinctly ascribed to the Princeton expeditions or to Scott either *in posse* or *in esse*—not that Scott failed to reject this grand scheme of Ameghino's but that virtually no well-informed paleontologist or other evolutionary biologist had ever accepted it. No doubt this was what Ameghino had in mind when he expressed the expectation that Scott's visit and study of the Ameghino collection would, in effect, force the paleontologists of Europe and America to admit the validity of all of Ameghino's work, even this heterodoxy. As mentioned also in Chapter 5 and here put in the words of Scott's autobiography:

> Ameghino had a theory, so passionately held that it amounted almost to an obsession, that all the various kinds of mammals, even including

> Man, had originated in Argentina and had spread from that region as a centre. He thus endeavoured to derive all the animals of North America and Europe, Africa and Asia from the fossil ancestors which had been discovered in Patagonia, an utterly inconceivable hypothesis. . . . I [Scott] have kept, as a souvenir, some of the sheets of foolscap paper, on which Ameghino made notes and diagrams of his genealogical trees. Aside from this idiosyncrasy, Ameghino had done an immense quantity of good and valuable work. [The curious reader will find some of Ameghino's "genealogical trees" in *Les Formations Sédimentaires,* cited above in Chapter 5.]

One cannot help feeling somewhat relieved that Florentino Ameghino died (in 1911) before the Princeton Reports modified so much of his "good and valuable work." They did accept and add to much of it, but in the end—that is, in Scott's summing up published in 1932—the theory "so passionately held" was practically neglected, without even the honor of an argument.

Scott was seventy-four years old when the tremendous Patagonian project ended with its final publication. He had nominally retired two years earlier, but he had fifteen more active years to live. It was thirty-two years between Scott's taking on the study of Patagonian fossils, then editing all the Reports of the Princeton University Expeditions to Patagonia and writing many of them, and finally publishing his overall summing up of his interpretation of the relevant geology of the Santa Cruz and of its mammalian fauna. During all that time, in addition to his academic duties, considerable travel, and frequent attendance at meetings, especially those of the American Philosophical Society, Scott was also writing books. One was a college textbook of historical geology, which ran through several editions, but most important here and for widespread paleontological knowledge was his remarkable *A History of Land Mammals in the Western Hemisphere,* first published by the Macmillan Company in 1913. This book did not presuppose any technical knowledge by the reader, although it was evidently written for reasonably sophisticated adults. On this subject Scott somewhat stiffly ended his preface as follows:

> While this book is primarily intended for the lay reader, I cannot but hope that it may also be of service to many zoologists who have been unable to keep abreast of the flood of palaeontological discovery and yet wish to learn something of its more significant results. How far I have succeeded in a most difficult task must be left to the judgment of such readers.

(I do not know what the general judgment was in terms of sales, and aged eleven I was too young for this book when it first appeared, but when I finally encountered it in college it was a godsend and greatly influenced the subsequent course of my life. I am not here going to review my own collecting and study of South American fossil mammals, which I have written sufficiently about elsewhere, but I trust that the reader of the present book will forgive this personal interpolation in connection with Scott's book.)

The first seven chapters of the book are introductions in a general way to methods geological and paleontological (or palaeontological, as Scott always spelled it), to classification of mammals as a whole and introduction to their anatomy in the parts usually found as fossils, to biogeography and the paleogeography of the Western Hemisphere, and to faunal succession in the two Americas. Then follow ten chapters summarizing the major groups (orders) of mammals then known as fossils in North and South America. The book is embellished and the text is supplemented and clarified by a large number of illustrations by R. Bruce Horsfall, the very able scientific illustrator who also drew many of the illustrations embodied in the Patagonian reports. A particularly interesting feature is that many of the extinct animals are represented as they probably were in life, the whole animals if the whole fossil skeleton is known, or just the heads if the rest of the skeleton is too imperfectly or not at all known. There are also many photographs of living North and South American mammals.

The last chapter is on "Modes of Mammalian Evolution," and this is a convenient place to discuss quite briefly Scott's attitude toward evolution at that time. The most important point is that from the start of his studies and throughout the rest of his life Scott held to the view that evolution is a fact supported by all relevant, material evidence and controverted by none. As to its "modes" he made four main points in this book: (1) Evolution is a highly complex process in which divergent, parallel, and convergent development is normally involved. (2) Development among mammals is usually by reduction in number of parts, but the contrary can occur; the law of the "irreversibility of evolution" often holds good, but its general validity is doubtful. (3) "Development among the mammals would appear to be a direct and unwavering process." (4) Whether evolution is gradual or sudden (saltatory) is undecided, but "there is no reason to support that the amount and rate of modification were always constant."

Scott discussed various possibilities for explaining these modes of evolution or finding their causes, but left these undecided. It is curious, in retrospect, that in this chapter he did not refer to Darwinian natural selection, then an old point of dispute, or to Mendelian genetics, a newer and then especially lively subject.

In 1937 Scott published a new edition of that book, under the same title but entirely rewritten and expanded by almost a hundred pages. A great deal of new knowledge had been achieved in the interval between editions—almost a quarter of a century—and with the advice and help of colleagues

Scott took excellent advantage of this. One simple example can be adduced to signalize the change. In the first edition *Astrapotherium,* the strange big beast that Moreno discovered and whose naming was in effect burglarized by Burmeister, was illustrated only by a sketchy restoration of the head, the skull being then all that was known. In the second edition there is a photograph of a complete skeleton lying as it was found by E. S. Riggs (see Chapter 10), a line drawing of the reconstructed skeleton standing erect, and a life restoration by Bruce Horsfall of a small herd of the animals in a likely environment.

Except for such advances in discovery, the rewritten book followed almost the same sequence and subjects as the first edition. The final chapter gives even more than before the impression that Scott had adhered to the bare fact of evolution and also but less firmly to some of its modes as suggested by the fossil record. However, he did here briefly discuss "Darwin's theory of natural selection," and firmly rejected it, surprisingly using as evidence *against* it such convergent evolution as between the sabertooth marsupial *Thylacosmilus* (another discovery by Riggs) and placental sabertooths, which most paleontologists take to be inexplicable *except* by natural selection. He also now mentioned, barely in passing, the rise of Mendelian genetics but saw no clear connection between it and the modes of mammalian evolution.

On the last point there are two relevant passages in Scott's autobiography as published in 1937. He relates there that in 1877, during his youthful exploration of the West, he had met an ancient character, "Uncle" Jack Robinson who when anyone told him anything would say, "Well, mebbe it is, but I don't believe it." Later in the autobiography Scott mentions the "development of the new science of genetics" and adds: "As for myself, I had witnessed the rise, culmination and decline of so many promising theories that my attitude was sceptical. Uncle Jack Robinson's formula, 'Mebbe it is, but I don't believe it,' is often very applicable." Scott still firmly believed in evolution, but in his last years he had virtually abandoned any hope of learning its mechanisms or causes.

Scott had retired in 1930, just fifty years after he had joined the Princeton faculty. He soon launched into three major efforts: complete revisions of his geology text and of the history of land mammals and also a complete monograph (published in sections, 1936–1941) of the known fossil mammals of the White River Oligocene, which includes the Big Bad Lands of South Dakota and extends also into Nebraska and Colorado.

The close friendship between Scott and Osborn, begun when they were undergraduates at Princeton, continued while both were alive. They were so often together at meetings and other events that they came to be considered as a sort of twin act. When they were studying in Heidelberg in 1881, a young Russian named Davidoff gave a farewell *Kneipe* (a German beer party) for which he decided to separate the twin act, addressing his invitation to Scott as "Sir W. Scott (ohne Osborn)." The "Sir" was evidently kidding (*Spass* in German) but "ohne Osborn" ("without Osborn") was probably a deliberate although amusing snub. This trivial incident was never forgotten by Scott, and it is brought up here because when Osborn died, in 1935, Scott sadly wrote, "it must henceforth and to the end be 'Scott ohne Osborn.' "

Scott found the "immense enterprise" of the White River monograph "too much for [his] unassisted strength." Glenn Jepsen (previously mentioned) assisted with several parts, and another vertebrate paleontologist who had been trained at Princeton, Albert Wood, wrote the sections on rodents and rabbits. Although Scott was eighty-three years old when the last section of that monograph appeared, he proceeded to write another, shorter monograph on a little known fossil mammalian fauna from northern Utah. This, his last publication, appeared in 1945. Bruce Horsfall, Scott's illustrator from the beginning of the Patagonia reports, continued so to the end of these last publications.

The venerable American Philosophical Society made grants to Scott for work on his final monographs and published them in its quarto *Transactions*. That society played a large part in Scott's life, and he also played a large part in the Society. He was its president from 1918 to 1925. This may be considered the highest honor that can be conferred upon an American scientist—as was noted in Chapter 1 of this book, Thomas Jefferson valued his election to this office as a greater honor than his almost simultaneous election to the presidency of the United States. While Scott was president the Society held monthly meetings; now it has an annual general meeting and only occasional other meetings. Scott was able to write that, during his presidency, "It was my remarkable good fortune, that I was never compelled, by ill health or stress or weather, to fail in attendance at a meeting, though, on one occasion, only the President and one of the Secretaries were present." (Princeton is not far from Philadelphia, the home of the Society.) In later years Scott was also present at most meetings until, for him, the last. In Philosophical Hall, on Independence Square, where the Society has met regularly for about two centuries, one of the meeting rooms is embellished by an oil painting of Scott in his elegant Oxford Sc.D. robe.

Scott was also a frequent visitor to the American Museum of Natural History, and those visits continued and for a time were even more frequent after Osborn's death. Much of the work on his last two monographs was done there, mostly in the library of the department of vertebrate paleontology, known as the Osborn Library, as it was established largely by deposit of Osborn's personal collection of books and papers. As he aged, Scott not infrequently fell asleep briefly over his work, and a worried woman member of the department used to check from time to time to be sure he was still breathing.

When in New York, Scott also frequently lunched in the American

Museum's staff restaurant, where he would regale the others at table with anecdotes of his youth and his many experiences since. The favorite, both to his audience and to him, dated from his never forgotten first expedition to the wild west. He was out afoot on his own one day, scouring the countryside for traces of fossils, when he found himself surrounded by a herd of cattle. Never having seen a man who wasn't on horseback, the cattle curiously began to close in on him, and he felt that they might soon crush him to death. He doubled up his fists with the thumbs sticking out and put his hands on each side of his forehead so that the thumbs resembled horns. He then rushed at the encroaching cattle, which quickly dispersed in the face of this amazing display. Scott's audiences never tired of his dramatic reperformance of this ruse.

Scott was dignified but amusing, staunch in his convictions but lovable. This abbreviated account may close with another one of the innumerable anecdotes both by or, as in this case, about him. At one of the meetings of the American Philosophical Society a member was reading a long, boring paper. Scott, in the audience, seemed safely asleep throughout the reading, but when comments were in order, he woke up and remarked (as nearly as the startled audience there could remember), "That was well done—exactly as I heard it here ten years ago."

Scott died of a heart attack on 29 march 1947 at the age of eighty-nine. His wife and their four daughters survived him.

Notes

For the account of Scott I have drawn especially from his conservatively but interestingly written autobiography *Some Memories of a Palaeontologist,* published by the Princeton University Press in 1939 when Scott was only seventy-one years old. I have also had access to the manuscript "Notes for an Autobiography," Chapter XXIX of which covered his days in Argentina. This manuscript was written for his children and then used in abbreviated and revised form as the basis of the published autobiography. As noted in the foregoing text there are some passages in the manuscript that were self-censored from the publication. Scott's other publications, especially the closing part of the long Patagonian Reports and the two editions of *A History of Land Mammals in the Western Hemisphere,* were also consulted again for the present book. For Scott's last years some memorial notices are available, and I have also drawn on my own memories of him, especially at the American Museum and to less extent at meetings of the American Philosophical Society, which I began to attend only about two years before Scott died.

William John Sinclair enters into the present book to a lesser extent. The essential outline and his life is given in some obits and in such reference works as *World Who's Who in Science*. I was also acquainted with him, but even more briefly and not as well as with Scott. Sinclair rarely visited the American Museum while I was there, but Glenn Lowell Jepsen, his favorite and my close friend kept us in touch.

INTERLUDE II

Botet

Francisco Javier Muñiz and José Rodrigo Botet are given places in this book because each of them had an interesting connection with the subject of fossil South American mammals, but each is here treated as a brief interlude because he did not personally make a special and definite contribution to scientific knowledge of this subject. Muñiz has been included because he is honored in Argentina as the first native of that country to collect fossils there to any noteworthy extent. He was also in other respects an unusual and interesting character. Botet is interesting because he brought together one of the largest collections of Pampean fossil mammals from Argentina and had it placed where one might not expect to find it. The collection is very impressive, but its scientific contributions to paleontology have been slight.

The birth of José Rodrigo Botet is on record with unusual precision as having occurred in Manises, Spain, at 5 P.M. on 2 April 1842. His parents were Onofre Rodrigo Aliaga and Teresa Botet Royo, who were potters in that town. In 1856 they went to live in Valencia, where they established a crockery and ceramic store. The still young José was aided in studies by his parents, and in about 1870 he followed the example of many poor but ambitious Spaniards and Italians: he emigrated to Argentina. There, eager to have a brilliant future, he studied in the university and was graduated as an engineer. That was an ideal time for a capable engineer in the blooming city of Buenos Aires and the development of the whole region.

Daró Rocha, then governor of Buenos Aires, was establishing the satellite city of La Plata, a few miles southeast of Buenos Aires down the shore of the Río de La Plata, and the plans for the city were drawn up by Botet. He continued with regional work on docks, bridges, and various other engineering and sanitary works in Buenos Aires and directed the extension of inland railroads and waterways. These activities made him rich, and he was also honored as cofounder (with Rocha) of La Plata.

His taste for adventure and his reluctance to settle down impelled him to go to Brazil, where he also worked with great success as an engineer. He directed the hydraulic power works on the Rio Tieté and constructed the electrical system of São Paulo, which was destined to become the largest city in South America. As a famous engineer he was elected president of the Paulista (São Paulo) Mining Commission, but he was not particularly interested in mining, and he returned to Buenos Aires. There he set up as "Rodrigo Botet and Company," devoted to the canal and port construction of La Campana.

In the course of excavations and other works in the great pampa of Argentina along and south of the mouth of the Río de La Plata, Botet became fascinated by the quantities of fossil bones that turned up. With a reasonable amount of time and a considerable amount of money, he determined to make the largest possible collection of such antiquities. To this end he employed Enrique de Carles, an Argentine collector and preparator who had become skillful in these activities while working for the Buenos Aires museum. It appears that Carles was the principal discoverer and excavator of the fossils, but often with Botet's enthusiastic accompaniment. Botet also purchased unusually good specimens which had been collected for sale as a sideline by various gauchos and estancieros.

As his collection grew, Botet was also made offers for parts or all of it, reportedly including one from the British Museum, but he turned them all down. He had another destination for his treasured bones. He had nostalgia for the city where he had grown up, and doubtless he also aspired to be considered there as a benefactor and a great man. With some difficulty he obtained a permit from the Argentine government to export the whole collection. He wrote to the alcalde (mayor) of Valencia offering to present the collection to the city, and received approval of that plan. In 1889 Botet and the collection sailed to Barcelona, where they were met by a commission from Valencia which escorted them back to that city by train. The great gift to the city was received there on 11 August 1889 and was housed first in the Faculty of Sciences of the University of Valencia. It was later transferred to the "Almadén" or "Alhóndiga," a curious and historic building which is now the Museo Paleontológico Municipal.

Enrique de Carles accompanied Botet and the collection to Valencia, and there he unpacked the fossils and began the long task of preparing them for study and for exhibition. However, "different circumstances," not specified, obliged him to return to Argentina some months later. After this point there is little more to say about Botet himself, although his name is still honored in Valencia and perpetuated in the museum that was effectually founded by and for him. In 1890, the year after returning to Valencia with his collection, he liquidated the funds earned and invested in Argentina. In 1904 he established in Valencia a company for exporting fruit. Little seems to be recorded about his last days, but it is hinted that they were unhappy. He moved to Madrid and died on 3 July 1915 at the age of seventy-three. Valencia honored his devotion by having his body brought back there in 1920 and ceremonially reburied in a mausoleum erected for that purpose by the municipal government.

After Carles left, the care of the Botet collection was taken over by successive members of the noteworthy Valencian family Boscá. First of this lineage as regards the museum was Eduardo Boscá Casanoves, who was born in Valencia in 1843 and died there in 1928. He became best known as a herpetologist, but was extraordinarily versatile, being also a doctor of medicine, a veterinary, a botanist, a mycologist, and ultimately a paleontologist. After he retired as a professor in the University of Valencia and took over the Botet collection he visited Argentine museums and also museums in Paris, London, Amsterdam, and Brussels, about which he published a book. He transcribed the field notes by Carles, sorted the fossils out, assembled some of the more complete specimens, and arranged exhibits—tasks to which he mainly devoted twenty-two years. In 1909 he published a small (16 page) "Catalogue-Guide" of the Valencia museum and in 1911, 1914, 1921, and 1923 he added publications on various groups well represented in the collection there.

Eduardo Boscá Casanoves married Dona Amparo Seytre Orios, and their older son, Antimo Boscá Seytre, was born in Valencia in 1874 and died there in 1950. Although not quite as polymath as his father, he was expert in geology and zoology. His father had taken him on visits to European and Argentine natural history museums, and he took many photographs, some exhibited in Valencia's museum. Antimo Boscá became director of that museum after Eduardo Boscá, and was in turn followed as director by Francisco Beltrán Bigorra, born in 1886, who had been a student of Eduardo Boscá. Beltrán was killed in an automobile accident in 1962, and the directorship of the museum was vacant for a couple of years and then was filled by Manuel Martel San Gil, who left Valencia for another city and university. Then the third of the Boscás in line, Fernando Boscá Berga, born in 1905, son of Antimo Boscá and grandson of Eduardo Boscá, took over.

What all this has been leading up to is the publication in 1964 of an extended catalogue of the Pampean fossil mammals in the Botet Collection. (There are a few other, older, specimens in the collection.) This was written by the then director, Martel, in collaboration with Emiliano Aguirre, a learned paleontologist, and still now, as I write this, a professor in Madrid. After 18 introductory pages and three unnumbered photograph plates, the catalogue has 99 pages of text followed by 119 photographic plates of specimens. The collection is surely, as claimed, the best representation of fossil mammals from the Argentina Pampean in Europe and, given the presence of numerous complete skeletons and skulls, quite possibly the best in the world. The book not only catalogues every specimen but also discusses them usefully. Stress is placed on the need for further study, but I believe that not much more is really needed. Most of the specimens belong to known genera and species, but many are more complete and instructive than the types. The collection contains three type specimens of species named by Eduardo Boscá in 1917, 1921, and 1923. Oddly enough it also includes the type not only of a species but also of a genus named by Florentino Ameghino in 1889. Just how Ameghino got hold of one of Botet's specimens, studied it, and had two drawings made, or on the other hand just how Botet got one of Ameghino's specimens, remains a mystery.

In the Museo Paleontológico, where the Botet Collection has been since 1906, there is a bas-relief of the head of Rodrigo Botet that shows him as a round-headed, small-eared man, wearing glasses, somewhat bald but with a nobly luxurious mustache, curled up at the ends. In the background is a partial view of the skeleton of a South American sabertooth cat, of which this museum has what is probably the most nearly complete skeleton known, and above Botet and the cat there is a grand parade of glyptodonts, of which the museum has a number of splendid specimens.

In 1960 my wife and I paid a visit to Valencia, and of course we spent considerable time in this rich but little known museum. It is in such an odd and unexpected place that even many paleontologists are not aware it exists. Certainly tourists are not, and many Valencians have never entered it. While we were there the only other person we saw was a guard.

Notes

The main printed source for this interlude is the catalogue by Martel and Aguirre mentioned in it. Its full title is: *Catálogo Inventario de la Colección Botet de Mamíferos Fósiles Suramericanos del Museo Paleontológico Municipal*. The full names of the authors are: Manuel Martel San Gil and Emiliano Aguirre Enríquez.

It was published in 1964 by the Archivo Municipal de Valencia. In addition to that date the title page interestingly also gives the date as: XXV Años de Paz Española ("25 years of Spanish peace").

Kraglievich; Cabrera

Ángel Cabrera, who will be discussed later in this chapter, became a chauvinistic Argentine. With reference to the pioneers Alcide d'Orbigny and Charles Darwin, he declared: "These naturalists did absolutely nothing for the progress of Argentine science; their collections went to the museums of Europe, and their works appeared in Europe. Thus they belong to the period in which the country is studied by outsiders, but without even beginning its [Argentina's] own scientific establishment."

Cabrera considered such a native establishment to have begun with Hermann Burmeister and to have been consolidated by the Ameghinos. In fact Burmeister was also a European—German—who, as has been noted, was brought to the Museo Público de Buenos Aires by Domingo Sarmiento. Cabrera himself noted that Burmeister's own researches on Argentine fossils did little to advance Argentine science but that he provided means for later research by others. It is also fair to note that although the most important part of all the Ameghinos' collection did eventually go to a national museum Florentino had sold many of their fossils to European and to one American (Cope), as mentioned previously.

As director of the then newly founded Museo de La Plata in 1889 Francisco Moreno brought in successively Alcide Mercerat and Santiago Roth, both from Switzerland. The museums in Buenos Aires and La Plata became polarized and interacted as rivals and critics rather than cooperators. Mercerat's work was hardly more than attacking the work of the Ameghinos, and Moreno also was deeply involved in the antagonism. As has been noted (in Chapter 5), Roth made one important contribution to Argentine paleomammalogy, but otherwise was more interested in geology as such than in paleontology. After Florentino's death in 1911 and Carlos's retirement the most likely successor to the Ameghinos in Buenos Aires was Lucas Kraglievich. In La Plata, after Mercerat and Roth, their successor for a time was Ángel Cabrera. This chapter is mainly a brief account of the lives and careers of those two men, both of whom I knew personally.

Lucas Kraglievich was born on 3 August 1886 in Balcarce, a town on the outskirts of Mar del Plata in the southern part of the province of Buenos Aires (far from the capital city of that name). A biographer has remarked that his birth demonstrated his impetuousness, or more poetically his "need for light" (*necesidad de luz*). He left the uterus so rapidly that the local midwife did not arrive in time, and his father delivered him. The father was Nicolás Kraglievich, an immigrant from Dalmatia who had a bakery in Balcarce and also a station for an intercity stagecoach. Lucas's mother was Lucía Carmen Vodanovich, born in Argentina as a daughter of a Genovese and a Triestine immigrant.

Having been taught to read and write by his highly literate mother, Lucas was sent to Buenos Aires, where his grandparents lived, to continue his elementary, secondary, and university education. There in a private school he finished his baccalaureate, in that time and place about equivalent to graduation from senior high school, and then went on to study in the Faculty of Exact, Physical, and Natural Sciences in the University of Buenos Aires. His aim at first was to become a mechanical engineer and he was adept in mathematics, but his heart was not in it. The grandparents lived in a modest ranch on the edge of the pampa, and he took to gaucho ways, becoming an adept horseman and also a guitar player. He did the required military term in a training camp and was commissioned as a *subteniente* (approximately second lieutenant) but was never called to active service.

As a young man Lucas Kraglievich was unworldly, and he remained so throughout his short life, trusting in others, deeply engrossed and learned in what interested him but quite naive in most ways. An anecdote from this period in his life is illustrative. He freely threw about what little money he had, and when a friend advised him not to do so he replied, "What do you want me to do, brother? The money is just kicking about in my purse!"

Lucas lost interest in mechanical engineering and did not finish for a university degree. He was, as his biographer put it, seduced by paleontology and geology, specifically as exemplified by the Ameghino brothers. To continue the expression of his biographer, he decided to give himself a doctor's degree. In the Argentine term, he was an *autodidáctico*—self-taught—an expression usually taken as pejorative when involving a profession. In fact there was then and there no other possible way to become a professional paleontologist. He taught himself the osteology and general anatomy of mammals and also the elements of stratigraphy and general geology. For the latter he spent ten months in 1912–13 in field studies of the sedimentary formations in the territory of Chubut and northern Santa Cruz, following in the footsteps of Carlos Ameghino. To earn a living during the training years he taught as a "subprofessor" in secondary schools (*colegios,* which are not colleges in the English sense).

In 1914 Dr. Eduardo L. Holmberg, a well-known Argentine scientist, became acquainted with Lucas Kraglievich and was impressed by his intelligence and his enthusiasm for paleontology. Holmberg also introduced Lucas Kraglievich to Carlos Ameghino, and the two took to each other. They had closely similar interests and also some similarities in backgrounds and characters. With Ángel Gallardo, then director of what had been named the Museo Nacional de Historia Natural de Buenos Aires, arrangements were made for Kraglievich to work in the museum, and particularly on the Ameghino collection, as a student. Having gained Gallardo's and Ameghino's confidence, he was made an honorary associate (*adscripto*) of the section of geology and paleontology.

Carlos Ameghino followed Gallardo as director of the Buenos Aires museum, and in 1919 he gave Kraglievich a formal staff position as technical assistant in the section of geology and paleontology. As early as 1916 Kraglievich had published in a Buenos Aires newspaper (*La Nación*) a long comment on "The Theories of [Florentino] Ameghino," and in 1917 he published at greater length a criticism of a work by Mercerat. That and also some later publications were in effect defense of the Ameghinos against their critics in La Plata. One of the most interesting counterattacks by Kraglievich was on the "Importance of the Paleontological Investigations of [Florentino] Ameghino," published in 1920. In 1921 and 1925 Kraglievich was junior co-author of two papers with Carlos Ameghino. These were Carlos's last publications, although he lived for some years thereafter. The 1921 paper was on a supposedly new species of *Megatherium,* and that of 1925 was on an extinct genus of the raccoon family.

While his staff position was still that of technical assistant, Kraglievich became more and more the de facto interim head of the section of paleontology. There he labored long hours on cataloguing the collection of fossil vertebrates, most of them in the Ameghino collection and estimated to include some eleven thousand specimens. Also in the 1920s he undertook his own researches, especially on ground sloths, which were particularly abundant in the collections and in the most accessible collection localities. Besides the work on ground sloths in this period he researched and wrote several important studies of fossil rodents.

When Santiago Roth died in 1924 Luis M. Torres, then director of the Museo de La Plata, is said to have invited Kraglievich to take over the position formerly held by Roth. Also during or after two rather brief field trips in Uruguay some of the geologists in that country urged him to move to Montevideo and to initiate a paleontological program there. In 1927 he published in an Uruguayan newspaper (*La Mañana*) comments on Uruguayan paleontological studies. At that time, however, Kraglievich was more at home and had better opportunities for his work in the Buenos Aires museum, and for a few more years he remained there.

There is an anecdote, quite possibly true, that at a banquet celebrating Carlos Ameghino's appointment as director of the national museum of Buenos Aires, attended by Lucas Kraglievich among many other Argentine scientists, Don Carlos rose to offer a toast to him as the *future* director of that museum. In the light of later events it is possible, but I think it quite unlikely, that Kraglievich took this seriously. His eulogists and biographers were illogically contradictory. They praised his modesty and his devotion to science above more worldly goals, and then they also represented him as abused, underrated, and made miserable because he was not more materially rewarded. The director of a national museum requires both administrative and political talents, and Kraglievich not only lacked but also scorned both. He was devotedly the worker at his last, and honored as such. He was well known, not at all ignored, by his fellow scientists in Argentina and also in Europe and North America. The National Academy of Sciences in Buenos Aires gave him the Eduardo L. Holmberg Prize, the highest award in its hands, and the Argentine Society of Natural Sciences elected him president. The repeated statement that he was an unrecognized genius, a prophet unknown in his own country, simply does not ring true. One can empathize with him as a hard-working genius, but there was no reason to pity him.

People who did pity Kraglievich and reproach the powers found support for this in the governmental decree that followed Carlos Ameghino's resignation from the museum directorship because of his increasingly poor health. This decree promoted Martín Doello-Jurado, an invertebrate zoologist and paleontologist, from his official post of *naturalista viajero,* (literally "traveling naturalist" but in effect a curatorial staff position) to directorship of the same museum. The report of this decree went on to say: "In the position of *naturalista viajero,* was designated Señor Lucas Kraglievich, who for some years past has given his valuable assistance in the Section of Paleontology (Vertebrates) as a supernumerary, being consequently in charge of said paleontological collections."

In fact, Kraglievich was not being passed over but was being officially raised to a higher and more responsible position in his own field of interest and research. It is, then, astonishing to find that not long thereafter (toward the end of 1930) Kraglievich was presenting his resignation to the director, and doing so in insulting terms. His letter to Doello-Jurado started as follows:

> Being informed by newspaper versions of a ministerial resolution bearing upon the question of the Museum, by which you seem to be implicitly confirmed in the directorial functions of the same, I consider

> that I have no prospect favorable to true scientific investigation to expect under this situation.

He then went on to complain bitterly, not about any treatment to himself but about a younger friend, a self-taught paleontologist, having been deprived of an opportunity to continue his researches in the museum. He finally blasted:

> In consequence, I refuse to continue carrying out the duties which as nominally in charge of the Section of Vertebrate Paleontology I have been performing in the establishment; and on presenting this resignation I declare my hope that there may also come for my country a time when its men of science cease to be appraised as mere cogs in the administrative machinery and that they be granted the consideration that is logically due them for their high intellectual rank and for what they represent for the country, consideration which, on my part, I honor myself as having fully deserved already from abroad, where I will doubtless find prospects for continuing my scientific work under the necessary conditions.

At this time Kraglievich was forty-four years old, an accomplished and well-recognized scientist, but in character no man of the world and, outside of his science, still the boy who gave his money to his friends because it was there. Now he was laying his career on the line impulsively. Kraglievich had been innocently caught up in a staff revolt fomented by an astute egotist whose ambition exceeded his ability.

The ploy did not work. After that letter the director really had no choice but to accept Kraglievich's resignation. Kraglievich had no choice but to go into exile as he had threatened. He did continue in Buenos Aires for some time, but in 1931 he went to Uruguay and settled in Montevideo. His friends in Argentina gave him a farewell dinner, attended by the ailing Carlos Ameghino and the increasingly famous Ángel Cabrera, among many others. Kraglievich was already well known in Montevideo, where a museum of natural history had been established under the directorship of Dr. Garibaldi J. Devincenzi. There Kraglievich was made chief of the section of paleontology, which he had been de facto in Buenos Aires with much better collections and facilities. There was, at least, a fair collection of fossil mammals in Montevideo, made by a devoted amateur, one Alejandro C. Berro. Kraglievich also had notes and manuscripts under way when he left Buenos Aires. For the short time he was to be in Montevideo he continued to study and to publish prolifically.

Hardly a year after his self-exile in Montevideo Lucas Kraglievich became seriously ill. His brother Nicolás and two friends went over and brought him back to Buenos Aires on 6 March 1932. He survived just one week, dying on 13 March at the age of forty-five. The funeral elegies by noted fellow countrymen made it obvious that he was indeed honored in his own country.

Kraglievich's active research was done in only fifteen years, but the extent of it was prodigious. In those years he had published about seventy-five research papers, six more were in press when he died, and another dozen or so were far along in script and some ready for publication. The majority of his publications were descriptions and classifications of fossil mammals, mostly Argentine but some Uruguayan. Among his first publications, in 1920, was also a short but classical study of fossil birds, especially of the gigantic, predatory, carnivorous extinct group known as phororachoids.

Although Kraglievich's many descriptive papers continue to be cited frequently by paleontologists, two of his faunal and stratigraphic contributions have had even more lasting effects on the study of the history and chronology of mammals in South America. The first of these, published in 1930 and written in Spanish (as were all Kraglievich's publications), was on "The Friasian (*Freaseana*) formation of the Río Frías, Río Fénix, Laguna Blanca, etc., and its mammalian faunas." The localities named are in far western Patagonia and were not examined by Carlos Ameghino. This added another defined age to the sequence set up by the Ameghinos (see p. 90).

At least equally important is his monograph on "The Pliocene age of the faunas of Monte Hermoso and Chapadmalal, deduced from their comparison with those that preceded and followed." This went beyond what appears in its title. It cleared up most of the whole sequence of South American faunas from the Miocene Friasian onward through the Pleistocene Pampean. It also gave generic faunal lists, complete as far as then known, for all the ages involved, and it is still a basic authority. This major work was completed in manuscript when Kraglievich died and was not published until 1934, then as a separate monograph printed in Montevideo. The cost of printing came to 369 Uruguayan pesos, a considerable sum at that time. The greatest contribution was from an Argentine lady, Doña Carmen Bernachi de Diaz, in the form of a jewel or ornament (*alhaja*) which was given to Lucas's wife, Doña Francisca Kral de Kraglievich, for this purpose and sold for 150 pesos. The remainder was made up by twenty-three other persons, some Argentines but mostly Uruguayanos, in cash contributions of from 1 to 25 pesos each.

Although a great deal has been written about Lucas Kraglievich, I have found no other reference to his wife, née Francisca Kral. They had a son who for some time followed in his father's footsteps. He signed at times simply as Lucas Kraglievich, which was not necessarily confusing, as his father of the same name was dead before the son began to publish. At other times the son was called Lucas Jorge, Lucas J., Jorge Lucas, or Jorge L. Kraglievich. He worked almost entirely in the municipal museum of Mar del Plata, about which more is said in Chapter 13.

As Martín Doello-Jurado has entered into the present account, brief further mention of him is relevant here. He was born on 4 July 1884 in the interestingly named town of Gualeguaychú in Entre Ríos, northern Argentina. He studied aquatic biology and particularly mollusks, both living and fossil, and throughout his adult life he was a, or indeed the, leading South American student of those subjects in museum and laboratory as well as in the field (for instance in Patagonia). He worked first in the invertebrate section of the Museo Argentino de Ciencias Naturales "Bernardino Rivadavia" (as the Buenos Aires museum of natural history had been renamed) and was made director of that museum when Carlos Ameghino retired. He was also a professor in the University of Buenos Aires, and he founded and was first president of the Argentine Association of Natural Sciences. Although he had likewise eminent predecessors as director of the museum, it was he who persuaded the nation to move it from its previously cramped and ancient quarters during the 1930's into the large and excellent building where it still is. He was largely instrumental in making this one of the great natural history museums of the world. He retired as director in 1946 and died suddenly on 9 December 1948 at the age of sixty-two.

From the present point of view the most interesting of Doello-Jurado's many publications is one outside his special field of research but not outside the scope of his knowledge. This was written in French, on the subject of "The *Megatherium*. Individual Death and Phyletic Death," and was published in 1939 as an introduction to an issue of the French journal *La Presse Medicale* devoted to medical matters in Argentina. Doello-Jurado wrote of *Megatherium* and other extinct mammals in Argentina that had undergone "phyletic death." He then philosophically discussed relationships—similarities and differences—between extinction, as phyletic death, and the inevitable death of individual animals. He ends with: "In conclusion, on comparing these two kinds of biological facts, individual death and *phyletic death,* the cause of which escapes us, one does not know how to say whether we have enlarged the scope of our knowledge or that of our ignorance."

After Roth died in 1924 and Kraglievich preferred to remain in the Buenos Aires museum, the La Plata museum needed a research paleontologist. In 1925 this position was taken over by Ángel Cabrera, who in that year on invitation from La Plata moved definitively from Spain to Argentina. He had been born in Madrid on 19 February 1879 and by 1925 was a mature and well-known zoologist. His whole education up to and through the Central University of Madrid was in his native city. In 1902 he joined the staff of the National Museum of Natural Sciences, still in Madrid, where he eventually was in charge of recent mammals. He visited museums in England and France, and he made field and office studies of recent mammals not only in Spain but also in North Africa, notably Morocco. By 1925 he was the leading mammalogist in Spain, and his reputation was spreading in other countries. He had then already published seventeen books, according to one of his biographers, and around a hundred shorter papers (*opúsculos*), a number of them in prestigious American and English technical journals. He was also secretary general of the Spanish Royal Society (Real Sociedad) and had other important positions among Spanish scientists.

But in just that year, 1925, the government of Spain was reorganized, and the nominal premier made himself dictator. (He is generally known as Primo de Rivera, but his full name, with Spanish *penacho,* was Miguel Primo de Rivera y Orbaneja, Marqués de Estella.) So Cabrera took off for the relatively calm and unfettered Argentina. This was not quite from the frying pan into the fire. He did escape the dreadful civil war in Spain, which eventually substituted one dictator for another, but some years later he found himself in another style of dictatorship under Juan Domingo Perón and his first wife, Eva.

At least safely established in Argentina, Cabrera took over as head of the department of paleontology in the La Plata museum. He had hitherto had to do almost exclusively with recent mammals, but finding himself in charge of a large collection of fossil mammals he turned to studies of the fossil specimens at hand, a new profession for him. The first result was a brief note "On the food of the megatherium" published in Spain in 1926 and of course in Spanish; thereafter most of his works were published in Argentina. He never wrote in any language other than Spanish, and although his publications make clear that he could read technical English, French, and German, he never, as far as I know, spoke anything but the most elegant Castilian—which is noticeably different from the Spanish spoken in Argentina and somewhat confusingly called *Castellano* (with the *ll* pronounced *zh* as always in that dialect; most other South Americans pronounce it *y* and the real Castilians, like Cabrera, pronounce it *ly*).

Cabrera's studies ranged at length through the fossil vertebrates, mostly but not entirely the fossil mammals, of the collections under his charge. The first important work, also published in 1926 but by the La Plata museum, was on "The fossil cetaceans in the La Plata Museum." (In a citation to this in a later reference work the Spanish for cetaceans, which is *cetáceos* and means whales, dolphins, and their relatives, was somehow printed as *Crétáceos,* which is no known language but looks as if it meant "Cretaceous"—incidentally, there are no Cretaceous cetaceans.) His next important work, published in 1927, was a monograph, "Data for knowledge of the Argentine fossil dasyuroids." The dasyuroids are carnivorous Australian marsupials, which at that time were often united in the same group (superfamily or even family) as the South American fossil carnivorous marsupials. Since then most students of these fossils agree that the South American and Australian group had a primitive ancestor in common but that they went their two separate ways in radiating and evolving on the two continents.

Another important monograph, based mainly on close study of the extensive earlier literature and also on the collections accumulated in La Plata, was "A revision of the Argentine mastodonts," published in 1929. There were many shorter papers based on the La Plata collection and on specimens found by others and referred to Cabrera for identification and description. The collections were also notably increased by field work by employees of the La Plata museum. For example, collection of fossil mammals in the arroyo Chasicó, in the southwestern part of the province of Buenos Aires had been going on for years and was continued under Cabrera's direction, mainly by Antonio Castro, a preparator (laboratory worker) in the museum. Most of the Chasicó results were published jointly by Cabrera and Kraglievich, especially in a monograph published in 1931, just after Kraglievich's self-exile, and finally completed by publication of another section by Cabrera alone in 1937, after Kraglievich's death. It is pleasantly noteworthy that the enmity between vertebrate paleontologists in La Plata and Buenos Aires was replaced by friendly collaboration after Cabrera went to La Plata and before Kraglievich left Buenos Aires. The fauna that they described together is now considered late Miocene, about ten million years in age, and is typical of the mammal age Chasicoan.

The most important field work under Cabrera's direction, and with his personal leadership, was in Catamarca. The area was scantily populated, and the party camped out during these expeditions, November–December 1927, February–April 1929, and January–March 1930. As will be related in Chapter 10, Elmer S. Riggs had collected there for the Field Museum of Chicago before Cabrera, in 1926, and there have been some problems about the age and the naming of the faunas found there. Cabrera called them "Araucaniano" and placed them in the Pliocene. Now they are usually called Huayquerian (pronounced wy-kay-reé-an), found to be five to seven million years ago in age and more or less overlapping the transition from late Miocene to early Pliocene. Riggs and Bryan Patterson summarized the faunas in 1939, and they were also considered by Patterson and Rosendo Pascual in 1942, but there is not yet a general monograph of all the known specimens of that land mammal age.

Although he visited a number of other fossil localities, the three coordinated expeditions of 1927, 1929, 1930 were the most important led by Cabrera. He did not publish a running account of them, but he was meticulous in noting the names of the places where the party camped and of the museum laboratory men who worked with him at each camp. The first expedition camped at Puerta del Corral Quemado ("gate of the burned corral"), and Bernardo Eugui was the museum preparator who worked there with Cabrera. The second expedition worked again at Puerta del Corral Quemado, and also at La Ciénega ("the marsh") and Las Juntas ("the crossroads"). Here Antonio Castro, previously named here as a collector and preparator of the museum, worked with Cabrera, and he did so again on the third expedition, which again worked at Corral Quemado and also at Loma Negra ("black hill"), located in between the Cerro Negro ("black mountain") of Hualfin (apparently a native Indian name) and the San Fernando river. For all three of the expeditions and camps Cabrera named a local man, Don Juan Mendez of Fuerte Quemado ("burned fort"), "who besides cooperating very actively in the extraction of fossils, acted as a guide and as the boss of the personnel of [local] workmen and muleteers."

The collections from that region were exceptionally rich in good specimens of glyptodonts, the heavily armored extinct distant relatives of the still surviving armadillos. These had been previously named and classified on the basis of mere fragments, virtually unidentifiable, and although Cabrera did not monograph the whole of the collections from this region, he did publish extensively on the glyptodonts. In doing so, this generally amiable man severely criticized some of the paleontologists who had previously published on this group, including by name Burmeister, Moreno, Mercerat, Ameghino, and Castillano. Having castigated his predecessors in this field, Cabrera went on to say:

> I have abstained from here entering into considerations of a phylogenetic kind. To establish genealogical lines, and even taxonomic divisions of an evolutionary character, on the basis of isolated fragments of caudal tubes [the armor around the tail] or loose sections of carapaces, is something beyond my capacity, and I am not in the least ashamed to make this declaration, given that everyone can see that the results of this kind of pedantry [*lucubraciones*] are hardly convincing.

Cabrera also noted that Castro and Eugui, who had been companions in camp on these expeditions, took over the long and tedious task of cleaning and preparing the fossils in the workshop of the department, putting them in shape for study. He also mentioned, somewhat parenthetically, that on the second expedition his son Ángel L. Cabrera, then a student, was in charge of making collections of recent plants. It is noteworthy that the son went on to become as well known as a botanist as his father was as a zoologist.

Having been in Spain primarily a student of recent mammals, and having taken a new position in Argentina in charge of fossil vertebrates, mostly but not exclusively mammals, it was a natural transition for Cabrera to make studies not only of totally extinct groups, such as the glyptodonts, but also of groups with both fossil and living representatives in South America. One of the first of those studies was "On the fossil and recent camels of South America" (1932), and it was followed by some others such as "The living and extinct jaguars of South America" (1934).

As time went on he tended more and more to go back to research and compilation on recent mammals, now generally those of South America. In 1940 he published with José Yepes as co-author a semipopular work on the "South American mammals," meaning those still extant. In 1947, which not entirely coincidentally was essentially the beginning of the dictatorship of Juan and Eva Perón, he retired from his professorship and curatorial position at La Plata and thereafter he had virtually nothing to do with paleontology. Somewhat curiously he was also a professor in the Veterinary School of the University of Buenos Aires and continued in that position until 1957. His principal work there was a monograph on the (living) Argentine rodents of the family Caviidae, that is, the guinea pig family which in the wild forms is confined to South America. This was published by the Veterinary School of the Agronomic and Veterinary Faculty of the University of Buenos Aires.

Cabrera spent his last years compiling a complete "Catalogue of the [living] Mammals of South America," published by the Buenos Aires museum (the Museo Argentino), which had been for some years installed in its much grander (but less central) building. The first volume of this final work by Cabrera was published in 1947. The second and last volume was issued in 1960, shortly after Cabrera's death in that same year. This large and useful compilation is not descriptive or diagnostic but is devoted to the formal classification, the technical (neo-Linnean) nomenclature, and geographic distribution of each species and also subspecies if, as in most cases, subspecies names had been applied and published.

In his research Cabrera was a descriptive zoologist primarily, and beyond that most interested in nomenclature (he was for some years an active member of the International Commission of Zoological Nomenclature). He wrote few papers that were at all concerned with theoretical matters. Only "The origins of the Argentine fauna," published in 1929, dealt with historical biogeography in a few (fourteen) pages. The topic obviously requires extensive knowledge of the successive fossil faunas of Argentina and in fact of most of the world. This was written when Cabrera had an excellent knowledge of many recent faunas, but unfortunately before he had acquired adequate knowledge of the fossil faunas even as that knowledge stood at that time.

Cabrera's other primarily theoretical paper was published in 1932 and is entitled "The ecological incompatibility, an interesting biological law." After many examples, all among recent animals (mostly not mammalian), the law is stated thus (in Spanish, of course): "The kindred animal forms are ecologically incompatible, their incompatibility being so much greater, the nearer is their affinity." Re-expressed in terms not yet generally current in 1932, that means that the more nearly related two animal groups (or taxa) such as species, for example, are, the less likely they are to occur in the same place and under the same conditions (sympatricly). After expanding on this idea, Cabrera summarized:

> Aside from this relationship with the distinction and establishment of species, the law with which we are concerned has a direct connec-

tion, according to what I have already indicated, with the problem of the extinction of certain species. Perhaps, in certain instances, even complete groups have succumbed to competition. Even if it is not from this viewpoint, ecological compatibility merits as much attention on the part of the paleontologist as on that of one who is dedicated to investigation of the biology of recent forms.

A present day biologist or paleontologist would express this somewhat differently or perhaps find it too conditional, but there cannot be much doubt that Cabrera's "law" does play some part in evolution. It is a good example of rediscovery in the history of science. Essentially the same points were made by Darwin in *The Origin of Species*. Later, but before Cabrera, they were applied by W. D. Matthew, a vertebrate paleontologist, in connection with interpretation of the presence or absence of similar mammals in the same ancient fauna. Still later again, essentially the same idea, expressed in different terms, was reached by an experimental biologist G. F. Gause. In fact what has been called "Gause's law" or "principle" is practically the same as Cabrera's "law," which was surely unkown to Gause. Along with real discoveries in science there are many rediscoveries that only pass as new. This is not written to disparage Cabrera. The idea was original to him, and if he could reach this on his own, it seems rather a pity that he did not spend more of his life in thought about evolutionary theory.

Cabrera was an indefatigable writer in styles ranging from the most technical to the most popular. One of his eulogists numbered his strictly scientific publications as 218 and his books, mostly written for the general but literate public, as 27. Shorter articles for various newspapers and magazines in the Spanish language ran up to uncounted hundreds. Most of his technical papers and books were illustrated by himself. He was perhaps not a great artist from a severely esthetic point of view, but he was certainly a superb illustrator both as a draughtsman and as a painter. He made precise pen and ink drawings of both fossil bones and living animals. His interests as a paleontologist were confined to fossil vertebrates, from fishes to mammals, but as a zoologist his books ranged over the whole animal kingdom. He was especially interested in horses, as fossils but particularly as live animals, both wild and domestic. His book "Horses of America" (1945) is still a standard. His book on "The extinct mammals" (1929) is the only one devoted especially to fossil vertebrates and perhaps was written too soon after his transfer to the science of paleontology. Among his last books is one published in Barcelona in 1950 and entitled *Zoología Pintoresca*, a Spanish title that needs no translation into English. It treats all sorts of animals at some length, even including some of the tiny one-celled forms no longer classified in the animal kingdom. It has 369 text figures, by Cabrera except for a few photographs, and 12 plates, half in pencil and half in full color, all by Cabrera. Its 32 chapters include one on "Animals through time" with a few restorations of extinct vertebrates and a photograph of the mounted skeleton of a Patagonian dinosaur in the La Plata museum. This chapter has a short passage on evolution, including mention of Mendelian inheritance pictured in mice. All the following chapters treat living animals without reference to their ancestors or other extinct relatives.

The Cabrera family was pleasant and hospitable. Scott mentioned that when he was in La Plata the Argentinians in general seemed to have little social interaction. Other Americans have not found this to be so. Scott's impression must have had a linguistic basis. Except for Florentino Ameghino, Scott's friends and acquaintances in Argentina were almost all Germans or German-speaking Swiss, and for occasional relaxation he went to meetings of the Deutscher Verein (German Union) in Buenos Aires. He did not really become acquainted with the usually charming Argentinians of Latin (mostly Spanish or Italian) descent.

Ángel Cabrera was still working and writing, mainly on the second volume of his catalogue of South American living mammals, almost to the day of his death, which was 8 July 1960. He was then well into his eighty-first year.

It is not within the scope of this book to discuss the lives or even to name all those who were to some extent involved with South American fossil mammals before the present active generation. This may, however, be an appropriate place just to mention a few of those who began what might have been important careers in Argentine paleomammalogy but who either dropped out or despite their efforts did not become important leading figures in this field in the disturbed times during and shortly following the lives of Kraglievich and Cabrera. As early as 1912, the year following Florentino Ameghino's death, H. Bluntschi, a distinguished Swiss scientist, said after visiting Argentina that "It could but shock me that science occupies such a secondary place in those South American countries. It is understandable that applied sciences should be those that are there preferably cultivated . . . but it is strange that pure science has not yet met more appreciation."

Among those whose promise was evident but whose whole accomplishment within this field seems disappointing in retrospect, the name of Cayetano Rovereto comes first to mind. He published in 1914 a large monograph on "The Araucanian strata and their fossils," which goes far beyond its title and discusses faunas geologically from Miocene to Pleistocene and geographically from southern Buenos Aires province north to Catamarca, but he hardly followed up those investigations.

Alfredo Castellanos published effusively from 1916 to 1973 (he lived from 1893 to 1975), but much of his work was devoted to Florentino's mistaken ideas of the antiquity of man in Argentina, and the rest was for the most part rather haphazard notes on various fragments of fossil mammals.

Three students of Cabrera in La Plata have been listed as the first female paleontologists in Argentina and probably in all of South America: Dolores López Aranguren, Enriqueta Vinacci, and Andreina Bocchino. However, the only reference I can find to any publication by them is a 1930 monograph on Argentine fossil camels by López Aranguren. Lorenzo Parodi, an employee at the national museum in Buenos Aires, was involved in the disturbance of 1930–31 that caused Kraglievich to leave. In those years he published three papers, one of them as junior author with Kraglievich, but apparently he published nothing before or after those.

There would not be much point in naming or discussing all South Americans who in the first half of this century had something to do with fossil mammals but made no substantial contribution to this subject. Something should, however, be said about Carlos Rusconi (mentioned already in the notes to Chapter 4) as a special case.

In closing his last major publication at the age of sixty-eight, Rusconi remarked that he had left school knowing more or less how to read and write and scarcely the "four rules of arithmetic." At the time when he ended his childhood this was the highest achievement that could be counted on from the third grade of primary school. Yet few men have written as much as he did. He soon became fascinated with fossils and developed a lifelong determination to make himself world-famous as a paleontologist. In due course he became acquainted with Carlos Ameghino, Kraglievich, and Cabrera. They encouraged him, and although he was not formally a student of any of them they also instructed him to some extent. Although he later wrote that he had been collecting fossil mammals in and around Buenos Aires as early as 1918, his first paper was published in 1927, "On a humerus of *Arctotherium* [an extinct bear] and another of *Felis* [the typical genus of the cat family] from the Pampean formation."

Thereafter Rusconi was extremely prolific, publishing in many media, or privately, and in two short-lived journals started by himself, the *Boletín Paleontológico de Buenos Aires* and *Ameghinia* (not to be confused with *Ameghiniana,* the now long-continued journal of the Asociación Paleontológica Argentina). Most of his publications before 1947 were short notes on fossils in his own collection or submitted to him from various sources. The most substantial work of that period was on "The Argentine fossil species of peccaries (Tayassuidae) and their relationship with those of Brazil and North America," published by the national museum of Buenos Aires, where up to that year he was working more or less under the supervision of Kraglievich. He also worked over the mainly mammalian fauna of sands underlying the rich loess of the Pampean and called Ensenadéen by Florentino Ameghino (*Ensenadense* in Spanish), placed in the Pliocene by both Ameghino and Rusconi but now considered Pleistocene.

In 1930 Rusconi was involved in the internal problems of the Buenos Aires museum which led to Kraglievich's resignation and eventual self-exile to Uruguay. Thereafter Rusconi was not free to use the Ameghino Collection in that museum, but for some years be continued to gather and study fossils mainly in and around Buenos Aires. In 1937 he moved permanently to Mendoza, capital of the province of the same name. In that city there was a small provincial museum called Museo de Historia Natural "J. C. Moyano." In due course Rusconi was made director of that museum, which later was moved to another building and then more simply designated Museo de Historia Natural de Mendoza. Rusconi reorganized the museum into six departments: paleontology (vertebrate and invertebrate), zoology, ethnography and archaeology, anthropology, folklore and history, and geology and mineralogy. Rusconi made himself head of the first five departments named, and for each he was also the entire scientific staff. For the department of geology and mineralogy he named as head and staff an amiable Mendocino, Manuel Tellechea, who had retired to follow his hobby which in North American slang is known as rock-hounding.

In 1947 the provincial legislature of Mendoza authorized the establishment and support of a *Revista del Museo de Historia Natural de Mendoza,* under the control and editorship of Rusconi. The first issue had two articles by Rusconi on some anatomical details of the bones of prehispanic Indians in Mendoza, one on designs from prehispanic pottery in Mendoza, one on shrunken human heads made by the Jívaro Indians of Ecuador, and one on a femur of *Megatherium* found in the province of Mendoza. Tellechea had in the same issue a short but well-written article on uranium, its properties and uses, explicitly including atom bombs and the possible development of uranium mines in the Province of Mendoza and elsewhere in Argentina.

Thereafter almost all the articles in the *Revista* were by Rusconi and were generally devoted to paleontology and anthropology, both in the broadest possible scope of those sciences. The *Revista* also sometimes included summaries of Rusconi's papers published elsewhere. Of those mainly on fossil mammals the most extensive was one based on fossils from the sands called Puelchéen by Florentino Ameghino and Puelchense by Rusconi, who had gathered some fragmentary fossils from them in and around Buenos Aires before he left there and on later visits. This appeared in two extensive parts published in 1948 and 1949 by the Universidad Nacional del Litoral in the city of Rosario through the university's Institute of Physiography and Geology, of which Alfredo Castellanos was then the director. In this work most of the specimens were so fragmentary that they bring to mind Cabrera's previously quoted stricture against giving new names to inadequate specimens. (I would here include *Macrocavia simpsoni* Rusconi, based on one scrap of a lower jaw.)

The most important mammalian fossils in the Province of Mendoza were from the Divisadero Largo ("long view"), a hill almost within the city of Mendoza. These were mammals that turned out to be of late Eocene or early Oligocene age. Fossils were just discovered there in 1946 by a student, Olivo Chiotti, who made the locality known to José Luis Minoprio, a skilled medical doctor and, in the full French significance of the word *amateur,* an amateur of fossils. Minoprio showed Rusconi the fossil localities, and Minoprio and Tellechea collected some specimens there. In 1946 Minoprio published the first paper on a mammal from this formation, but this was not well understood, and knowledge of the fauna depends almost entirely on Minoprio's continued collecting and collaboration with North American paleontologists. The importance of this is that the fauna partly fills in a time gap or hiatus just before the Deseadan and can be considered either early Deseadan or as the type for another land mammal age and fauna, the Divisaderan (or Divisaderense). The point here in reference to Rusconi is that despite his energetic exploration and study he failed here to find or to understand such an important fauna that was for many years almost at his doorstep.

Writing on many different topics in Mendoza, Rusconi especially emphasized invertebrate fossils, mostly trilobites, very ancient and extinct distant relatives of the still extant crustaceans, and also at a wide remove everything having to do with the prehispanic Indians of that region. In both those general fields his work was strongly criticized by more experienced authorities and excited typical ego-defenses from Rusconi, but those are not the concern of the present book. Including this much about him here is because the extraordinary bulk of his work has given him considerable repute among some of his fellow countrymen not especially knowledgeable about the quality of that work.

As far as I know, the last publication by Rusconi was a large book—489 pages with 278 figures and followed by 46 plates—on "Extinct animals of Mendoza and Argentina," published in 1967 by the provincial government of Mendoza

Notes

A collection of the works of Kraglievich in three volumes was made under the direction of Alfredo J. Torcelli, who had compiled the life and works of Florentino Ameghino, and completed by Carlos A. Marelli. This was published in La Plata in 1940 by the Ministry of Public Works of the Province of Buenos Aires. The first volume has an introduction by Torcelli, centering somewhat bitterly on the break between Kraglievich and the Buenos Aires museum but also giving the essential data of his biography. The present writer also became well acquainted with him during his last months in Buenos Aires.

I was also well acquainted with Ángel Cabrera in the 1930s and by later correspondence. The second volume of his "Catalogue of the mammals of South America," published in 1961, begins with a short biographical notice by Jorge A. Crespo. Another useful obituary by Max Birabén was published in the journal *Neotropica* in 1960.

The abbreviated and not entirely sympathetic discussion of Carlos Rusconi is also based on personal acquaintance, both in Buenos Aires and in Mendoza, but more especially on his publications. He was highly self-centered, and this breaks through even when he was writing about other things. For example, in his book "Extinct animals of Mendoza and of Argentina," cited in the text above, the last pages under the title "Recapitulating" are devoted to a recital of his accomplishments and virtues, and to his bitterness that he had not been treated as well as he felt he deserved.

Especially useful for this chapter, and some others, has been the following summary of the progress of Argentine vertebrate paleontology up to the beginning of the 1960s:

Rosendo Pascual. 1961. Panorama Paleozoológico Argentino: Vertebrados. *Physis,* 22 (63): 85–103.

Riggs; Patterson

During the years 1922 to 1927 the North American paleontologist Elmer Samuel Riggs covered more South American territory geographically and a wider range stratigraphically than had been explored there by any previous collector of fossil mammals. Although he started his field work while Cabrera and Kraglievich were still alive and active, he fits into our historical sequence at this point because the main studies of his collection came later (essentially started by Bryan Patterson) and still cannot be considered complete.

Riggs was born in Nineveh, Indiana, on 23 January 1869. Little seems to be recorded about his early days, but it is known that he attended the University of Kansas as an undergraduate and graduate student, with a bachelor's and, in 1896, master's degree. For the following year, 1896–97, he was an assistant paleontologist in that university's paleontological museum, and then was a fellow at Princeton University in 1897–98. He went to the Field Museum of Natural History in 1899 and spent the rest of his working life at or on expeditions for that museum, starting as an assistant curator and advancing to associate curator in 1921 and curator in 1937. He retired in 1942, aged seventy-three. Thereafter, in 1943–1946, he was named as honorary curator at the University of Kansas, but he retired even from that honorary position in 1946, when he definitively ceased active work or studies and moved to Siloam Springs, Arkansas. He died in 1963 at the age of ninety-four.

While still a student in the 1890s Riggs worked with Barnum Brown collecting dinosaurs in Wyoming for the American Museum of Natural History, under the general direction of Henry Fairfield Osborn and the more immediate supervision of J. L. Wortman. As Riggs wrote much later, he and Brown used then to discuss at night what fossil collecting they might do later, including the possibility of collecting in Patagonia, the riches of which were just then being revealed by the Ameghinos. They did both carry out this dream, Brown much earlier than Riggs, as related here in Chapter 7, but more briefly and less importantly, as will appear in the course of the present chapter.

Riggs's early work when he went to the Field Museum was mainly the collecting and occasional description of dinosaurs from western United States, but he also did some work on western North American fossil mammals. In 1921 the idea of a South American expedition was put forward more definitely, and the eponym and principal supporter of the Field Museum, Marshall Field, expressed interest in this plan. The expedition was finally decided upon in July 1922, and the then director of the museum, Dwight Davies, forthwith wired to Riggs, who was then in Alberta, Canada, collecting dinosaurs with John B. Abbott and George F. Sternberg. They crated up their dinosaur bones and shipped them to Chicago, to which all three of them also hastened. With Riggs, now associate curator, in charge, these three would constitute the technical staff of the coming expedition. Abbott, of British origin, had long been at the Field Museum as a collector and laboratory preparator. Sternberg was an old hand in the business, having collected for several museums and also on his own, offering the specimens for sale.

There followed frenzied preparations, with gathering and packing of equipment, until on Monday, 6 November 1922, Riggs could write in his personal journal, which he kept up for much of the expedition: "The day of starting on our long-planned journey to South America has arrived. All of the rush and strain of preparation, of ceaseless work, days, nights, and Sundays has been leading up to this."

On the way by train, they stopped at Pittsburgh, where O. A. Peterson discussed Patagonia with them and gave them letters of introduction. Next they stopped in Washington, to visit the National Museum and the Argentine Embassy—the latter to get letters admitting their equipment into Argentina duty-free. Thence, on 10 November, to Princeton, where they met both Scott and Sinclair. Riggs worked late that day and most of the following day, particularly on getting information as to Hatcher's fossil localities. On 12 November they reached New York. They visited the American Museum of Natural History but apparently saw none of the staff there "on account of a flower show"—not otherwise explained.

On Wednesday, 15 November 1922, they finally sailed on the passenger vessel *Southern Cross*. As many of us have learned in writing journals or logs of journeys and expeditions, only after we are under way do we catch up with how we got there. On 17 November, two days out, Riggs got around to catching up on the last ten days in his journal. He also had been asked by his wife, Fannie, to keep a log, a daily record of the journey, and he commented in his journal, somewhat resignedly, "This may be worthwhile for her and the boys [his sons]."

On 28–29 November the ship stopped overnight in Rio de Janeiro. Unimpressed by this impressive city, Riggs remarked only on its being hot and stuffy and on the Portuguese language not being a medium of communication, for his party at least. On 2 December they put in at Montevideo, and Riggs went ashore briefly but had nothing more to write about it than that the muddy Río de La Plata "hardly justifies the name 'silver river'." (As commentator in the present book, I may point out that the Spanish explorers did not name this estuary as a "silver river" because the estuary itself was silvery [*argentine*], but with the hope that its tributary rivers led inland to lands with silver mines.) On 3 December Riggs and his party arrived in Buenos Aires, where they remained, except for one day in La Plata, until 22 December. In La Plata the party "were greatly impressed with the extent of the paleontological exhibits" and were amiably received by Luis María Torres, then director of the La Plata museum, and by Santiago Roth, then its resident vertebrate paleontologist.

On his visit to La Plata, Riggs was appalled to learn that the Argentine government had passed a new law about the collection and export of paleontological and related specimens. The law appeared virtually confiscatory, and it could have greatly impeded, if not altogether stopped, any further collecting by non-Argentinians. A commission had been established to apply the law and, as Riggs wrote to the director of the Field Museum, "to exact its pound of flesh." The nominal head of the commission was then Carlos Ameghino, as director of the Buenos Aires national museum, but Torres was acting head because Carlos Ameghino was then in poor health—although as will be recalled he lived some fourteen years thereafter. Torres reassured Riggs that the commission would not be rigid in this respect and that with considerable red tape, as is normal in almost any country, the Field Museum's collections could be exported practically intact, as indeed they finally were.

On 22 December the expedition was finally on its way to Patagonia, sailing on the government steamer *Asturiano*. (Asturias is a province in northwestern Spain, in the same general region as Galicia; Asturians and Galicians [*Gallegos*] were among the Spanish immigrants to Argentina, sometimes looked down on by Argentinians of Castilian descent.) Late December is early summer in Argentina, and this is sheep-shearing time in Patagonia, where almost all the landed proprietors raise sheep. The steerage of the *Asturiano* was crowded with shearers, mostly from Spain and Italy, who were temporary immigrants for the shearing season only.

On Christmas day the ship arrived in Puerto Madryn, the northernmost port of Patagonia. Riggs had little to say about the place except that it was a "struggling town of wooden buildings with corrugated iron roofs. A few were painted." He also noted the absence of trees and the prevalence of bushes, which he called "chico" or "chica"—a word that is not to my knowledge the name of any particular Patagonian bush but only means that they are small. One further journal entry mentions the clumsy carts on which baled wool is brought to a port. Riggs spent the afternoon sleeping, and went to a Christmas dance in the evening. Here he was impressed by the "cosmopolitan mixing of all classes," and by the fact that a "stout rural policeman" and a "fleshy matron" did a "patiga" dance together. (I never heard of a "patiga" and cannot find the word in dictionaries; Riggs may have been referring to a *zapateo,* which was a popular sort of tapping dance throughout Argentina.) All in all, Riggs's first day in Patagonia, to which he had been looking forward for years, does not seem to have excited him or even impressed him emotionally, as it has many visitors there from Darwin onward.

From Puerto Madryn the ship went on southward to Comodoro Rivadavia, more commonly called simply Comodoro, which just means "commodore" in the naval sense. It was the shipping point for a small oil field, not strictly speaking a port as it had no harbor or pier. Riggs had little to say about it, but later was to revisit it and ship fossils from there. Thence on, still to the south, to Puerto Deseado, where the party went ashore for a few hours and Riggs was impressed that "Automobiles of Ford, Overland or other makes were everywhere in the streets." Then to the next ports, San Julian and Santa Cruz, where the party did not land, and finally to the Río Gallegos, which was their destination. This was reached on the last day of the year, Sunday, 31 December 1922, nine days after leaving Buenos Aires, forty-six days after leaving New York, and fifty-five days after leaving Chicago.

Welcomed by English-speaking people in the town, capital of Santa Cruz province, the party engaged rooms in a somewhat primitive hotel, acquired an Italian translator from Spanish to English and vice-versa, were introduced by him to the provisional governor of the province, then a Colonel Isa, and were registered, fingerprinted, and measured politely by the rural police. This would undoubtedly have been done in any case, but Riggs wrote to his museum director, Davies, in Chicago that it was the result of there having been a recent rebellion in the province for which "some four hundred men were executed." He added, "Please do not communicate this to our families."

From the port, the party followed in the first footsteps of John Bell Hatcher in Patagonia, to an estancia with the Indian (Tehuelche) name Killik Aike, owned by the Felton family and now run by Don Carlos Felton. This family was one of many who had moved to this part of Patagonia from the Falkland Islands or were descendants of settlers who had done so. As in the Falklands they raised sheep here, and this way of making a living had spread to most of Patagonia, although in more central regions there were few large and rich estancias like that of the Feltons. Here most of the families of Falklands origin had remained bilingual and ethnically British although elsewhere in Patagonia Welsh and Boer immigrants were becoming Spanish-speaking ethnic Argentinians within a generation or two.

On the land of this estancia there were exposures of the Santa Cruz formation rich in fossil mammals. As related in Chapter 7, these had been collected profusely by Hatcher. It may be added that in 1896 Handel Tongue Martin, of the University of Kansas, had also made a collection here, but as no addition to knowledge of South American mammals has been made on the basis of this collection, there is no reason to discuss it here.

The Field Museum party set up a camp on the Felton estancia, not far from the Felton's home and other buildings, and at long last they could and did turn to on what they came for, fossil mammals—here the Santacrucian early Miocene fauna already well known by the successive and voluminous work of the Ameghinos and Hatcher, Scott, and Sinclair. The Chicagoans first worked for about three weeks in, or especially at the base of, cliffs on the north bank of the Río Gallegos (here, as Hatcher had found, being eroded by high tides from the sea). The party continued to collect in the Santa Cruz formation from early January 1923 through early May. From the Felton estancia they worked, as Hatcher had, east to Cabo Buen Tiempo ("Cape Fairweather") and then north to the mouth of the Río Coyle. As fossiliferous exposures are virtually continuous along and near the coast, for purposes of recording field localities Riggs divided them somewhat arbitrarily into eight stretches or areas, mostly designated by the name of an estancia or its owner. In succession these were Felton's, Halliday's, and Rudd's estancias, then La Angelina, La "Pousta" (probably a misspelling of Posta), Coy (or Coyle) inlet, and Lago "Sección Smith." Halliday, of "Halliday's Estancia," assigned one of his men, Hawkins, to Riggs's party as interpreter and cook. Hawkins proved to be far less than perfect as a cook, but Riggs wrote in his journal, "Hawkins of no more use than ever but I could not discharge him and leave us no interpreter." (The party that was baffled by Portuguese in Rio was for a time, at least, about equally baffled by Spanish in Argentina.)

They were traveling with a truck and a passenger car as well as saddle horses. The truck was nearly lost when it broke down on a beach and was repaired just in time to escape from the incoming tide. They sometimes were accommodated in a building on an estancia, for instance first in what Riggs called a "galfon" (correctly a *galpón,* the Argentinian word for a large shed or storehouse) and then in a cook-house or "cocino" (correctly *cocina*). Usually they camped in tents, and in the almost incessant Patagonian wind had trouble with the stakes being pulled out, a bother that must be expected by anyone who tries to tent in Patagonia.

At the end of their first month Riggs went back to Río Gallegos to get funds and supplies. The expected funds had not been sent from Chicago, and Riggs took out a personal loan from the local Anglo-Bank, with which he paid off Hawkins and bought supplies for a month. From Halliday's he took on E. A. Miller, an experienced mechanic, as a camp man and translator in place of Hawkins, and also rented Miller's car, a "Ford runabout," for transport additional to the truck and riding horses.

At the end of their work on the Santa Cruz formation's near-coastal exposures, the whole collection was assembled at Halliday's estancia, there packed with some difficulty, and then forwarded by ship from Río Gallegos to Buenos Aires. The list of specimens, sent by way of a friend to Luis María Torres as acting head of the Argentine commission, included 282 specimens of fossil mammals, among them 177 skulls and some skeletons more or less entire. Among these, as later prepared and studied in Chicago, the prizes were two skeletons so nearly complete that they could be restored, mounted in lifelike poses, and exhibited. Both are still unique exhibits on display in the Field Museum. One, *Homalodotherium,* was a herbivorous hoofed mammal belonging to the order of Notoungulata, named by Roth. The genus had been named by the zoologist Sir William Henry Flower in 1873 (more fully characterized in 1874) on the basis of scanty remains sent to the British Museum. (This was the only South American specimen published by Flower.) Riggs's essentially complete specimen was eventually described by him in 1937, twenty-four years after it was collected. As mounted, standing, the animal has a sturdy body suggestive of a rhinoceros, although without horns and not especially related to rhinoceroses.

The other, even more extraordinary skeleton is that of an *Astrapotherium.* As was noted here in Chapter 9, the basis for this genus was a partial specimen found by Romero but sneakingly named by Burmeister in 1879. The Field Museum skeleton, as described by Riggs in 1935, is mounted in a prone or prostrate position. It has enormous upper tusks and an odd skull that indicates either a trunk or an inflated nose and upper lip. It and its close relatives are so unlike any other mammals that it is generally classified in a separate order, Astrapotheria, named in 1894 by another English zoologist, Richard Lydekker. The Chicago specimen was also described at some length, with figures of the skeleton and life restorations by Bruce Horsfall, in the second (1937) edition of Scott's *A History of Land Mammals in the Western Hemisphere.*

With the Santa Cruz collection complete and started on the first leg of its long journey to Chicago, Riggs's next moves seem to indicate a degree of credulousness or naïveté surprising in a man so sober in character, so experienced, and so mature (he was now fifty-four years old). In Río Gallegos he had met a man named (or purported to be named) J. G. Wolfe. Wolfe now told Riggs of great discoveries in Patagonia, one of a very old, "fossilized human skull" and another of an "enchanted city." Wolfe now wished to join the party and to guide Riggs to these discoveries. The affair is made more curious by the fact that Davies, as director of the Field Museum, cabled instructions that Riggs employ Wolfe. Riggs was not wholly without doubts, but he did take Wolfe on.

On 27 April 1923, Riggs, Miller, and Wolfe set out for El Paso de Santa Cruz, a locality near the mouth of the Santa Cruz river where an English nurse, a Mrs. Vendrino, supposedly had the fossil human skull. On arrival there, Riggs was told that Mrs. Vendrino had become insane and had gone to Buenos Aires for treatment, taking the skull with her. Riggs did eventually track down the supposed skull in Buenos Aires and found it to be merely a somewhat curious stone. From El Paso de Santa Cruz, Wolfe guided Riggs by a difficult road and trail far inland to near Lake Cardiel, where Wolfe's "enchanted city" was said to be. The "city" turned out to be merely a lava dike, like many such in this part of Patagonia. Wolfe next told Riggs about an "ancient cemetery" of fossil mammals to which a man who claimed to have worked with Carlos Ameghino twenty-five years earlier would guide them. This "guide" merely led them in a circle, and Riggs finally lost faith in Wolfe. Disillusioned, he wrote in his journal, "Further inquiries were made in regard to Wolfe. He exhibited as qualifications a long angular personality, a bald head, a bland manner, a mode of speech which never said anything specific but always ended in an unfinished sentence." Having traveled several hundred miles to no avail, the party returned to Río Gallegos.

It was now May and the Southern Hemisphere winter was beginning. The party planned to go by land far north, into southern Buenos Aires province, and to winter in Bahía Blanca. They had gone only as far as Comodoro Rivadavia when the bad weather made any substantial further travel impossible. Principally to save hotel bills, Riggs sent Abbott and Sternberg a few miles north to camp near Pico Salamanca, not a fossil land mammal locality. Riggs went off by ship to Buenos Aires. The winter was severe, even for Patagonia, and during July Abbott and Sternberg "camped" comfortably with an oil company employee near Comodoro. They managed to collect some fossil invertebrates and parts of a fossil whale from the Patagonia Formation, which outcrops near and almost in Comodoro. Finally, as winter waned in late August, they went to the locality known as Cabeza Blanca, of which more later.

In the meantime Riggs had been more comfortably housed physically but harassed in other ways in Buenos Aires and La Plata. Torres, still acting head of the Argentine commission, insisted that his whole collection be brought to La Plata for inspection. Kraglievich came down from Buenos Aires to look after the interests of the national museum there, Carlos Ameghino still being indisposed. Eventually all the specimens were checked against Riggs's field book, two unidentified ones being taken to Buenos Aires by Kraglievich but later returned to Riggs. Riggs waited in Buenos

Aires week after week; finally, on 1 September, seven weeks after he had arrived there, Torres gave him a letter on behalf of the commission authorizing the export of the collection subject to further approval by the Minister of Justice and Education. This being promised, Riggs arranged for shipping the collection to the United States, also arranged for his nineteen-year-old son Harold to come to meet him in Argentina, and then on 7 September took the steamship *Buenos Aires* to Comodoro.

Arriving in Comodoro on 15 September, Riggs found that Abbott and Sternberg had left that vicinity about three weeks earlier and were now encamped on the Río Chico. He engaged a member of the Boer colony settled in that area as a guide and set out in search of his colleagues, only to be overtaken by Sternberg, who had gone to Comodoro by another road and was on his way back to camp. On the next day Riggs rejoined Abbott and Sternberg at Cabeza Blanca (or Loma Blanca). This fossil locality was found by Carlos Ameghino in 1894 and was then rich in specimens of what the Ameghinos called the *Pyrotherium* fauna and beds, now called Deseadan following Albert Gaudry, as mentioned in Chapter 6. The Amherst Expedition under Loomis had collected there in 1911, as I have previously noted. Although all the people in that region knew it as Cabeza, or Loma, Blanca, Riggs called it "Loomis Hill." That gave far too much credit to Loomis, and Riggs's renaming of it has not been adopted either locally or among paleontologists internationally. It was Carlos Ameghino, not Loomis, who found this fossil deposit and mapped it, as later published by Florentino Ameghino in 1906. With some repetition, I must also say in this context that Loomis made a mess of the local stratigraphy and an even worse mess of his publication on the Deseadan fauna. The Field Museum party did not find such rich picking left, but in a few days they did take out twenty-nine specimens of Deseadan mammals.

Riggs then went back to Comodoro, sending instructions for Abbott and Sternberg to meet him there. Somehow those instructions went astray, and for over a week Abbott stayed idle at the estancia of a Sr. La Grange, a local settler, while Sternberg and La Grange himself prospected for fossils without success.

It was now late in October and the spring rains were unusually heavy, making inland travel difficult and at times impossible. The party did get together in Comodoro and on 31 October set out for the vicinity of Lake Colhué-Huapí and the adjacent village Colonia Sarmiento. They crossed the high Pampa Castillo, so called there but topographically better designated as a *meseta* or, in English, plateau. From there they went down into the Valle Hermoso ("pretty valley"—something of a misnomer) and to the south bank of the Río Chico, well upstream from Cabeza Blanca and only a couple of miles from its source in Lake Colhué-Huapí. (Río Chico means simply "small river," and since Argentina is replete with small rivers, this one is distinguished as the Río Chico del Chubut because when water runs in it, which is not all the time, it is a tributary of the large and perennial Río Chubut.)

The beds that Riggs came upon here were much older than the mammal-bearing beds at Cabeza Blanca, and they contained dinosaurs. Some unknown person had piled up dinosaur bones here, and then evidently just gone off and left them. The Field Museum party spent about a week collecting dinosaur bones. Then they remembered that they were after fossil mammals and that their objective in this area was the Great Barranca south of Lake Colhué-Huapí, long famous among paleontologists. As related here in Chapter 4, this unique feature was first discovered as a fossil-collecting area by Carlos Ameghino, who eventually found that it uniquely contains fossil mammals of four distinct faunas and ages, one above another in a single continuous cross section. (It was not determined by the Ameghinos or by the Field Museum collectors that a fifth distinct fauna lies below those four.) The Field Museum party spent a few weeks there with considerable success, although the fossils at any level are not as numerous or as well preserved as those in the best collecting areas of the Santa Cruz formation, all of later age than those in the Great Barranca. Three short excursions to the south turned up a few more fossils, all of Deseadan age like those from Cabeza Blanca and from high up in the Great Barranca. The Deseadan is not there at the very top, where the fauna and land mammal age are appropriately called Colhuehuapian. (Geologically speaking they are of the latest Oligocene or earliest Miocene epoch.)

Elmer Riggs's son Harold joined the party at the Great Barranca. Determined to increase mobility, Riggs went in by train to Comodoro, a track between Comodoro and Colonia Sarmiento having been laid. (It is still in existence but no longer in use.) He engaged a German cook, Otto Defarode, and bought a car in which he was driven back to camp. When they finished work at the Great Barranca, they made a reconnaissance westward to the head of the Río Senguer, which feeds the two lakes, Musters and Colhué-Huapí, but found no fossil mammals.

They spent Christmas in Colonia Sarmiento, which had one English-speaking inhabitant. Thereafter they explored the Cerro San Bernardo, which is west of Lake Musters, and found only a few isolated dinosaur bones. Then they worked northward from Colonia Sarmiento and found a few fossil mammals considered by Riggs to be Casamayoran in age, called the *Notostylops* fauna by the Ameghinos and considered Cretaceous in age but now known to be early Eocene. Santiago Roth had worked in that general area, but most of his fossils found thereabouts were somewhat later in age. The Field Museum party then returned to Sarmiento and again went west to the region around the Cerro San Bernardo. At about this time Riggs wrote in his journal: "The heat of summer was on, water was bad. One after another, men were taken ill. A dismal January 23rd commemorated my birthday and 55 years of varied experiences."

Now they found more dinosaur bones, and they were joined by the well-known fossil-reptile specialist Professor Doctor Friedrich Freiherr von Huene from the University of Tübingen, Germany. He had been collecting with considerable success in Brazil, but his visit to Argentina was relatively brief and not very productive.

Since they went up to Comodoro Rivadavia the Field Museum party had been collecting fossils in the southern part of the Province of Chubut, in central Patagonia. At this point Riggs was not satisfied with their results there, although it would later be found that some of the fossils collected were excellent and important. Now early in March 1924, with the southern autumn approaching, Riggs decided to go south to the province of Santa Cruz, not for further work in the Santa Cruz formation, from which they had already sent back a splendid collection to Chicago, but to investigate localities with older faunas from which classic collections had been made by Carlos Ameghino and by Tournouër. After some scouting on the way, they headed for Punta Casamayor and not far from there for what the Ameghinos had called Cañadón Tournouër. Riggs assumed that the early mammalian fauna known as Casamayoran, following Gaudry, would be well represented near Punta Casamayor.

For several days they searched the region along the sourthern shore of the gulf of San Jorge between Punta Casamayor and Puerto Mazaredo. In fact Carlos Ameghino had found only one unidentified fossil mammal in that region, and although Tournouër had made a small Casamayoran collection there it was not identified and published until after Rigg's visit. The Field Museum expedition found a few relatively unimportant specimens and went on their way, first to Puerto Deseado and hence southwest to an estancia called La Flecha ("the arrow"). Here they spent a month, guided by Tournouër's *vaquero* ("cowboy" or in this usage a local guide) to the place where Tournouër had camped in 1903. Here they made a good collection from the *Pyrotherium* beds or Deseadan, including *Pyrotherium* itself. While the others worked there Riggs went northwest to Jaramillo and beyond to Pico Truncado, where a good deposit with Deseadan fossils was located. When the party was through at La Flecha it moved there. They then went on south to the Cerro Cuadrado, finding a few fossil mammals on the way but chiefly seeking (and finding) a locality from which petrified pine cones and wood were known to have come. The party made a moderate collection of the smaller pieces, and these were eventually identified and described in 1935 by a paleobotanist, George Rieber Wieland (1865–1953) at Yale University. This site was later made a national monument by Argentina and is considered one of the two most celebrated petrified forests, the other being in the Painted Desert National Monument in Arizona,

U.S.A. Curiously the petrified trees in both these forests, so distant from each other, are closely allied to *Araucaria*, a genus of conifers still living in the Southern but not in the Northern Hemisphere.

By this time it was late fall and turning cold. The fossils had been shipped to La Plata for examination by the commission, and Riggs and the other men headed north separately over bad roads and in unpleasant weather. Arrangements had been made for Abbott and Sternberg to return to the United States for the duration of the southern winter. They had now been in the field for a year and a half, and they were worn out and anxious to see their families before returning to continue the expedition. As he traveled north Riggs rceived mail in Bahía Blanca, including a cable from the director of the Field Museum stating tersely: "Expedition must complete South American work before return. No appropriation available for extra ocean trips." In fact Abbott and Sternberg were already on a ship headed for the United States. Riggs wrote a bitter account of this development in his journal, ending with: "For their sakes I hoped the men would not obey my directions. But a wireless sent to recall them." The expected message sent by them from the ship was, "Impossible to return." In fact neither Abbott nor Sternberg ever did return to South America.

Riggs and his son took a train north to Buenos Aires. The new collections had arrived there, but some were being held in the customs (*aduana*). While waiting, Elmer and Harold Riggs went up to Paraná, in far northern Argentina, where some fossil mammals had been found. They did not acquire any and returned to Buenos Aires and to the examination of the commission, now headed by Doello-Jurado but including Carlos Ameghino. There were the usual difficulties, especially because some specimens were in plaster jackets and removal of these would probably damage them. At this point, according to Riggs, Carlos Ameghino said: "These specimens cannot be laid open to inspection. It would be a crime to sacrifice the specimens under such circumstances." He then assured Riggs that "the matter had been carried out in a way sufficiently candid to reassure everyone concerned." Presumably an interpreter was present. A compromise was made and almost all the specimens were exported.

Riggs had applied through the American consul in La Paz, Bolivia, for permission to collect rather late (Pleistocene) fossil mammals in the Valley of Tarija. Such permission being given, the Riggses, father and son, in July 1924 went by train to the Argentine-Bolivian border and then overland by bus to Tarija. The fossil bones, presumed to be of giants, had been noticed in this valley by early explorers. Natives of the valley had collected many of the fossil bones, and an ex-senator residing there, Luis Echezu, had made and purchased a large collection. Part of this was purchased from Echezu in 1903 by Comte Créqui-Montfort of Paris and presented to the Paris natural history museum where Marcellin Boule was in charge of fossil vertebrates. With the assistance of Armand Thévenin, Boule wrote and had published in 1920 a large memoir based on this collection. Riggs was familiar with this, and he was ambitious to obtain a similar collection for the Field Museum. With local assistants, who became somewhat recalcitrant, the Riggses did amass 126 specimens, especially of glyptodonts. These were successfully shipped to Chicago by way of Buenos Aires and New Orleans. They include some striking specimens, but no noteworthy addition to knowledge of the fauna has been based on them.

Riggs and his son returned to Buenos Aires on 20 December 1924, and Mrs. Riggs joined them to celebrate Christmas there. Riggs was ordered by Davies to return to Chicago as soon as arrangements to ship all collections and to store equipment had been made. In his somewhat recalcitrant way Riggs did make several more relatively short trips in Argentina, but he finally left in late January, going to Chicago by way of Europe, where he used the time of accumulated vacations to visit museums. On 23 April 1925 he finally did arrive home again with wife and son. He had spent more than two years in the field in South America.

The instructions to store equipment in Argentina implied that the Field Museum would send a second expedition there, and as he returned from Europe Riggs began planning to get this under way for the following year. As neither Abbott nor Sternberg would be available, the first problem was to find at least one replacement for them. After searching and consideration, Robert C. Thorne of Vernal, Utah, was selected. He had been in the army in World War I (the "Kaiser's war"), and, as Riggs wrote, he was "camp-bred, physically strong, accustomed to out-of-door life, and resourceful." Like many people in Vernal, which is near Dinosaur National Monument and some other fossil fields, he had also an interest in fossils, but he was not a professional in that field. As a second man, Professor von Huene volunteered the services of a former student at Tübingen, Rudolf Stahlecker, who was an accomplished stratigrapher and collector, having worked in the field with von Huene. He was also a veteran of World War I. Thus Riggs was taking into the field two men who had fought in the same war but on opposite sides. As might have been expected, they took a violent dislike for each other, and although they stayed on for most of the expedition this did not make for congenial camping together.

It had been known for some time that fossils were abundant in parts of the province of Catamarca, in northwestern Argentina, and Riggs decided to go there first. With Thorne he left the United States on 10 April 1926 and arrived in Buenos Aires on 28 April. Riggs had to tend to some business and supplies, and Stahlecker had not yet arrived from Germany, but a month or so later the whole team was established in camp at an Indian village called Chiquimil. Through a Colonel Wieser, who was working as an archaeologist in this general region, two Catamarcans, Juan and Felipe Mendez, were employed to assist in camp and in collecting.

A rich deposit of fossils near Chiquimil was worked by the party until near the end of June. These were mostly fossil mammals, but the most striking specimen found there was a gigantic flightless, carnivorous bird. Riggs wrote of this: "A great fossil bird, new to this formation and probably new to science is the last discovery." This was a phorohacoid, and it was indeed new, but it was not adequately described and named until 1960. Stahlecker made sections of the rocks on which the stratigraphic positions of fossils found could be recorded, while Thorne and Felipe Mendez generally located specimens and Riggs and Juan Mendez excavated them. This locality is in the Santa María valley, where members of the party eventually prospected for fossils for some fifty miles. Colonel Wieser informed them that he had seen fossils some hundred miles to the southwest near the town of Puerta del Corral Quemado (where Cabrera also camped, later), and the camp was moved to there. Although considerable collecting had been done in the vicinity of Chiquimil, none was known to have been done around Puerta del Corral Quemado. There was field evidence of three successive faunas there, covering a long span in time between the Santa Cruz fauna, far to the south, and the Tarija fauna to the north.

For inspection of the collection for export, the Argentine commission sent Cabrera to the camp, and he found that checking the field records was adequate. Apart from the big bird at Chiquimil the most unusual discovery of the expedition was at Puerta del Corral Quemado, where they found several partial specimens of a large sabertooth marsupial, a remarkable example of convergence between a marsupial and the independently evolved placental sabertooth cats, notably *Smilodon*. Cabrera approved taking all these specimens to Chicago for study, and this became one of the few South American fossils that Riggs later described on his own, published briefly in 1933 and fully in 1934. He gave it the new generic name *Thylacosmilus,* which carries the meaning of "sabertooth marsupial." (It will be recalled from Chapter 9 that Cabrera subsequently collected extensively from the Catamarca localities for the La Plata museum.)

Assembling, boxing, and transporting the massive collections on muleback and by ox-cart to the railroad was a difficult problem, but was solved. Riggs, Thorne, and Stahlecker then went to Buenos Aires. A license for further collecting was somewhat belatedly granted, and Riggs decided to turn now to the later faunas in the province of Buenos Aires. He sent Thorne and Stahlecker to Necochea, a small port east of Bahía Blanca and southwest of Mar del Plata on the southern coast of Buenos Aires province. He went to Bahía Blanca to get the truck and camp equipment that had been stored there since June of 1924, and then joined the others at Necochea. From January into early May they collected faunas Pampean in a broad sense, with greatest success by following up the Río Quequén from its mouth to some sixty miles upstream, still in the southernmost part of Buenos Aires province. They also visitied the long famous locality Monte Hermoso, but found little there. It had been frequented—one is tempted to say infested—by fossil-hunters for about a century from Darwin onward. Incidentally, along the Quequén Salado Stahlecker found about half the skeleton of a *Smilodon,* the placental sabertooth that invaded South from North America and ecologically replaced the native South American marsupial sabertooth.

The party returned to Buenos Aires in early May 1927, and there Stahlecker left to go back to Germany. Eventually he joined the Nazi party, and there seems to be no record of his later life. Here, however, I may briefly resort again to the first person. In 1930, when I was planning my first expedition to South America, von Huene wrote to me suggesting that the expedition be made jointly between the American Museum of Natural History and the University of Tübingen, with Stahlecker representing the latter. Fortunately for me, as I believe, this turned out to be impracticable. And so Stahlecker vanishes from the scene.

Riggs, still fossil-hungry, decided to go back to Bolivia. In Tarija, not very surprisingly, a revolution was under way and the province was under martial law. Beyond the area of the strife there was another smaller valley, that of Padacayá, and the party worked there with great success. Exceptionally, this valley apparently had not been worked over for fossils. The prize, found by Thorne, was an unusually complete skeleton of *Megatherium,* supposedly but perhaps questionably of a species distinct from the Luján specimen with which this book began. Later they could return to the valley of Tarija. There, Riggs tried to purchase the tremendous private collection of Luis Echezu, a part of which, as will be recalled, was purchased earlier for the Paris museum and was monographed by Boule and Thévenin. Echezu would not at this time sell the whole collection, but Riggs did purchase 6000 bolivianos worth. (I find no record of the value of a boliviano at that time.)

At long last Riggs left South America forever, journeying to Chicago from Mollendo, in Perú, by way of Panama and New Orleans. He and Thorne had been away for over eighteen months.

Here a new character must be introduced, one who contrasts rather sharply with Riggs in temperament: Bryan Patterson. There can be no derogation of Riggs's remarkable achievements, but he was rather pedestrian even in his persistence and devotion to his profession. Patterson, or Pat, as all his friends called him (only his wife called him Bryan) was less predictable, but no less devoted to his work, and I found him more interesting and more congenial.

Bryan Patterson was born in London, England, on 10 March 1909. His father was Lieutenant Colonel John Henry Patterson, D.S.O., and his mother Frances Helena (Gray) Patterson, LL.D. Both parents were distinguished.

His mother earned the first law degree ever granted to a woman in Great Britain. His father, as a member of the Royal Engineers, had been engaged in planning and supervising the construction of a railroad in Kenya from the port of Mombasa to the inland capital Nairobi. (It was later continued northwest to the Uganda border.) He is probably best remembered for his classic book *The Man-Eaters of Tsavo,* published in 1907. In the vicinity of Tsavo, now a National Park, lions killed a great many of the laborers building the railway, and Colonel Patterson eventually killed the lions. They were later mounted as museum displays in the Field Museum of Natural History, a fact that influenced Bryan Patterson's career in an oddly indirect way.

The son and only child of this redoubtable couple was educated at home by his mother until he was fourteen. He then entered Malvern College, an institution of no great antiquity (founded in 1862). Although considered "important" it was without the prestige of Eton, for example. It was not a college as a unit of a university but a "public school" in the English sense of the words, that is, a pre-university school that we would call a "preparatory school" or "prep" in current American slang. Pat and Malvern did not find each other congenial, and their ways parted. Pat eventually had the distinction of becoming an internationally renowned scientist and Harvard professor without ever having been graduated from a school or holding an earned degree. (He did become a Harvard M.S., but this is an unearned degree automatically given to full professors of science at Harvard, and, it has no real significance.) Pat was an autodidact and was evidence that being self-taught can work very well if one is a good teacher.

When Pat left Malvern, Lieutenant Colonel Patterson found himself with a not particularly controllable seventeen-year-old son on his hands. He solved the problem through his acquaintance with the president of the Field Museum of Natural History, who agreed to see that the rather intractable youth was given some occupation in the museum that required no training. In 1926 Pat came to the United States and to the Field Museum. The great South American collections of fossil mammals made by Riggs had not yet been adequately prepared in the laboratory, so Pat was assigned to Riggs as an assistant, to learn about fossils the hard way.

As one of his fellow laboratory men later recalled, one day Riggs sent Pat to the library to check a reference, and thereafter Pat was off and running, or one might better say "off and self-teaching." In 1931 he published his first paper, which was not based on the South American collection but on a specimen that Riggs had collected in South Dakota in 1898. Patterson made this the type of a new species of alligators named by him *Allognathosuches riggsi.* From that time to the end of his life his publications were prolific, and he left a great deal of still unpublished manuscript materials when he died.

In 1932 Pat began collecting fossil mammals for the Field Museum, first in western Colorado where in 1933, 1937, 1939, 1941, and 1947 he and members of his party made large collections of previously litle known or unknown Paleocene and Eocene faunas. His skill in finding, collecting, preparing, and studying fossils, mostly fossil mammals, developed quickly. In 1937 he was moved from the laboratory to the curatorial staff of the Field Museum, where he was promoted to curator of vertebrate paleontology before he left the museum in 1955.

In 1934 Pat had married Bernice Maurine Caine, known as "Bea" to all friends. They had one child, a son Alan Patterson. Pat became a naturalized citizen of the United States in 1938, and in 1944 he enlisted as a private and went overseas in the 16th Regiment, First Infantry Divison of the A.U.S., serving in Europe in the Second (Hitler's) World War. He was engaged in the Normandy invasion and was captured by the Germans in the Battle of the Bulge. He escaped twice, but was twice recaptured and finally set free by the advancing American forces. Two legends from the war period, probably true but possibly not, give some insight on Pat and also his relationship with his father. One is that when Pat was reported as Missing In Action, Lieutenant Colonel Patterson is said to have exclaimed, "The damn fool probably stepped on a land mine." The other is that Pat enlisted as a private instead of going into a (commissioned) Officers Training Camp because he considered enlistment more patriotic in his adopted country. Pat denied this after the war, but it could well have been in his sometimes oddly retiring—and at other times brash—character.

Until considerably later in his career Pat had himself collected fossils only in the United States, but he had become more and more involved with the Riggs Collections from South America. Although he was still concerned with North American research, in 1939 an important study of stratigraphy and faunas (or "faunæ" as the title called them) of the collections by Riggs's parties in Catamarca appeared in *Physis,* an Argentine natural sciences journal. This was published with Riggs as senior and Patterson as junior author, but it is fairly obvious that the stratigraphy was due to Stahlecker and the faunal summaries to Patterson.

In 1934 to 1937 Patterson published detailed and important papers, mostly on notoungulates, based on preparation and study of specimens in the Riggs Collections. In this same period Riggs had published on the three most striking, if not necessarily most important scientifically, of his discoveries in Argentina: the marsupial sabertooth *Thylacosmilus* in 1934 and the extraordinarily near-complete skeletons *Astrapotherium* (published in 1935) and *Homalodotherium* (1937). Patterson had a hand in preparation of these specimens, and there is reason to think that he also had a significant part in their study. (Riggs acknowledged his assistance, but not coauthorship.) In 1935 a large paper nominally by Riggs and Patterson was published on some

Eocene (Casamayoran) specimens collected by Riggs's party, mostly in the Great Barranca. There is little doubt that most of this was written by Patterson, under Riggs's supervision.

Here I must again comment in the first person. In 1934 I had completed my own early field work in Patagonia, and the preparation and study of the resulting collections was well under way. Patterson, on behalf of the Field Museum, and I, on behalf of the American Museum, agreed that I would undertake the revision of the Eocene faunas as then known—Casamayoran and Mustersan—and that Patterson would revise the following Deseadan and Colhuehuapian faunas. Riggs insisted that Patterson should study and describe the best Casamayoran specimens in Riggs's collection before lending them to the American Museum for my study. Patterson reluctantly agreed, and Riggs arranged publication with himself as nominal senior author. It was petty of me, but I then retaliated by first describing a few of the best of the American Museum Deseadan and Colhuehuapian specimens before lending them to the Field Museum.

In the 1950s Patterson had two Guggenheim Fellowships for study abroad, and he spent them in Argentina, mostly in the two great museums in Buenos Aires and La Plata but also in the smaller, but excellent, municipal museum in Mar del Plata. He studied and made full notes and, where needed, drawings of all the types and other important specimens of the Deseadan and Colhuehuapian faunas, and some others, in all available Argentine collections. He later used some of these notes for publications, the most significant being: on Deseadan rodents with Albert Elmer Wood (1960), on marsupials with Larry Marshall (1978), and on general review of South American fossil mammals with Rosendo Pascual (1968). He also made the notes available to other students (including me), but he did not revise the Deseadan and Colhuehuapian faunas as a whole, and most of his notes are still unpublished.

In his last years in Chicago besides his Field Museum work Pat had an adjunct connection with the University of Chicago, where he collaborated with Everett Claire Olson in the development of an interdisciplinary program in paleontology. He thus had some academic background when he went to Cambridge, Massachusetts, as an Alexander Agassiz Professor in the Museum of Comparative Zoology. As that is an untenured position and nominally not a teaching one, he was also made an unpaid but tenured professor in Harvard University, and he taught graudate students, eight of whom obtained doctorates under his tutelage. His own field work in 1956 to 1964 and much of his research were devoted to North American fossil mammals, but in 1963 through 1967 he led large expeditions to northern Kenya. Pat himself did little research on the resulting collections, but they have been a resource for colleagues and students, and they were followed up by others who have made much-publicized discoveries broadly in eastern Africa.

Until 1958 Pat had done no field work in South America, but in that year with Al (Alfred Sherwood) Romer of the Museum of Comparative Zoology and Rosendo Pascual of the Museo de La Plata he was a member of a joint expedition of the two museums to northern Argentina. The main purpose was to collect early Mesozoic reptiles, but Pat did have a brief opportunity to visit the Divisadero Largo formation near the city of Mendoza, and he found some rather fragmentary mammals there. That formation has a peculiar fossil fauna with mammals unknown elsewhere. What is still the definitive work on that fauna was published in 1962 jointly by Pat, José Luis Minoprio, and me. In our opinion that fauna could be considered as Deseadan but probably somewhat earlier than the typical Deseadan in Patagonia. However Pascual has named a distinct land mammal age Divisaderan as preceding the Deseadan.

In 1970 Pat's nontenure professorship in the Museum of Comparative Zoology was not renewed, but he retained and was now paid for his tenured professorship in Harvard University and continued to work in the museum. Also in 1970 he was asked by the International Executive Corps to go to the town of Estanzuela in Guatemala, where fossil mammals had been found, and to report on possible provision for a local museum to display them. Under his direction a considerable collection was made and identified as an interesting late Cenozoic mixture of mammals of South and North American origin involved in the Great Interchange of land faunas when the two formerly separate continents were joined by the Isthmus of Panama as a land bridge. The collection was retained in Estanzuela in a museum named the Museo de Paleontología "Bryan Patterson." As far as I know a description of the collection has not been published, and I am also unsure of how that Museum has fared in the later and present political and warlike activities in Guatemala.

Later in the 1970s Pat led three separate expeditions in South America in areas outside of Argentina and Bolivia. In the regions then explored by Pat few finds of pre-Pleistocene mammals had been made. In 1972 the party went to Venezuela, in 1973 with different personnel to the Central Amazon Basin in Brazil, and in 1974 again with different personnel to the upper Amazon basin in eastern Peru. In Pat's own opinion these expeditions were unsuccessful in the search for fossil mammals, and he did not publish on the few mammalian bits that were found. These expeditions did collect some well-preserved fossil reptiles, but Pat also did not publish on these.

Having mentioned some of the things that Pat worked on but did not publish, I should add that throughout his working life he was studying and publishing importantly on fossil mammals, some collected by himself, mostly in western United States but also in Florida, and some studied in various museums, mostly in the United States, England, and continental Europe. As an example, in Texas Pat made the first important collection of early (by some stratigraphers designated as "middle") Cretaceous mammals. In 1956 he published an extremely important and germinal study of one group of these. That is just one of a large number of Pat's studies and publications that are not relevant to the theme of the present book but that helped lead to wide recognition of his contributions to science, for example to his election as president of the Society of Vertebrate Paleontology and as a member of the National Academy of Sciences.

In 1975 Pat retired definitively and ceased to teach, but for a couple of years he continued to publish important studies. He died of cancer in Boston on 1 December 1979 at the age of seventy.

As a person Pat was not only a devoted scientist on the job but also what is known in vernacular American English as a "character" with what one of his colleagues has called a "vibrant personality." He was at some times and about some matters reserved and withdrawn, but at other times and in other circumstances he was rollicking and even downright eccentric. In this respect he was the sort of person about whom anecdotes circulate and are long remembered. My own close friendship with him began well before he left the Field Museum and became even closer when we were both at the Museum of Comparative Zoology from 1959 to 1970. Without being too extensively anecdotal, I will mention just three of the occasions that illustrate the most "vibrant" side of his personality. At one of the meetings of the Society of Vertebrate Paleontology in Canada my wife and I invited Pat and some other members to have a drink in our hotel room after the afternoon session. When we got to our locked room, Pat was already there serving drinks: he had climbed into the room through the transom. At another meeting—this one in Atlantic City—he somehow found a bicycle of the sort that had an enormous front wheel with the saddle on top and a tiny following wheel. This kind of bicycle, a "penny-farthing" in English slang, went out with the nineteenth century. Somehow Pat got on board and then cycled all around the banquet room. On still another occasion, on a field trip, we were having a meal in a small town quick food restaurant that had a moving belt carrying used crockery out to the scullery. Pat climbed on board, lay down on the belt, and thus left the party. These and many other occasions exemplify a side of Bryan Patterson not evident in his publications or in the usual written record.

Notes

The account of Riggs is based on all of his publications, including nontechnical articles in the *Field Museum News,* and also nontechnical articles by Larry G. Marshall in the *Bulletin of the Field Museum*. Riggs kept an unpublished personal journal of his expeditions in South America, and Larry Marshall has made copies of extensive parts of these available to me. I was also personally acquainted with Riggs for some years.

As evident in the preceding text I was more closely and longer acquainted with Bryan Patterson, and also with his wife and son, who survive him. I have read all his published work, and much of the unpublished. Word of him frequently appeared in the *News Bulletin of the Society of Vertebrate Paleontology*. I have also had in hand somewhat detailed obituaries by Farish A. Jenkins, Jr., of the Museum of Comparative Zoology, and William D. Turnbull of the Field Museum of Natural History. There are innumerable anecdotes of his antics when playful in mood, and I hope that someone will sometime assemble a collection of them, without neglecting his serious side nor delving untactfully into his private side.

Stirton

For years some pioneering geologists in South America believed that there was a continuous connection between the Atlantic and the Pacific Oceans across the whole continent from what is now the Amazon Basin in Brazil to what is now the Gulf of Guayaquil between southwestern Ecuador and extreme northern Peru. Others doubted this, and as will be seen in this chapter it has been disproved (although in 1983 an advertisement for a ship tour up the Amazon still stated it as factual). Something was soon known about the land mammals of northern South America during the Pleistocene; that is, the last million years or so. That increasing knowledge indicated beyond serious doubt that the northern and southern faunas within South America were then much alike, differing only locally, as they still do today. However, it still seemed possible, even if not likely, that a cross-continent water barrier had earlier separated the two regions. To test this hypothesis it was necessary to find definitely pre-Pleistocene faunas in northern South America. These could then be compared with the ones, fairly well known by 1905 or so, from southern South America, especially Argentina.

This crucial evidence, long desired and believed to be available, was not definitively obtained until expeditions led by the American paleomammalogist R. A. Stirton with the assistance of the government of Colombia, were carried out in that country in 1944–45, 1946, 1949, and 1951.

Ruben Arthur Stirton, always called "Stirt" by colleagues and other friends, was born on a farm near Muscotah, Kansas, on 20 August 1901. Although he became a highly educated and distinguished scientist, something of his rural youth hung on in later life, sometimes in his speech and in his informal manner. One of his teachers spoke of him as "a diamond in the rough." He was a gem, figuratively, and he was in the rough inasmuch as he did a great deal of field work under difficult conditions. When young he was greatly interested in natural history, and in his twenties he went to the

University of Kansas. There he majored in mammalogy and ornithology and was graduated with an A.B. in 1925. From 1922 to 1925 he was an assistant in the Dyche Museum of that university, and in his senior year he also taught classes in ornithology. In 1925 to 1927 he went as a (Recent) mammalogist on the Donald R. Dickey Expeditions to San Salvador in Central America. In 1927 he started in graduate school at the University of California (Berkeley). There he was first an assistant to William Diller Matthew, who had gone to California in 1927 from the American Museum of Natural History. Matthew turned Stirt's interest from Recent mammals to fossil ones. He was made a curator of fossil mammals in the University of California's Museum of Paleontology in 1928. Having attained his A.M. in 1931 and Ph.D. in 1940, he was made associate professor in the university in 1946 then in 1949 director of the museum, professor of paleontology, and chairman of the department of paleontology. (His professor and predecessor in that position, Matthew, had died in 1930 while Stirt was still a graduate student.)

From 1939 to 1944 Stirt was following the usual course of a North American paleomammalogist: collecting and studying fossil mammals of western United States. Of his forty-four technical papers published in that period, the most influential was probably his "Phylogeny of North American Equidae," a history of the whole horse family on this continent, published in 1940. Now, more than forty years later, it is still a classic and well worth restudying.

In 1944, as noted above, his attention turned to South America, more specifically to Colombia. Prior to that he had married Lillian Miller and they had a son, Jack. On 12 August 1944, bound for Colombia, Stirt, Lillian, and Jack sailed from San Pedro, California, on the Argentine boat *Rio de La Plata*. While in harbor in Acapulco on 17 August the ship caught fire, burned all the Stirtons' equipment and clothes other than what they had on, and sank two days later. On 22 August the family went to Mexico City, and two days later Stirt sent Lillian and Jack back home. On 27 August he started on by plane to Bogotá, Colombia, with stops in Guatemala and Panama. He finally reached Bogotá on 30 August.

For this expedition Stirt had support from a Guggenheim Fellowship, and in Colombia he was further aided by the Colombian government's Servicio Geológico Nacional and by geologists and oil companies operating there. Among the former were the Colombian geologist José Royo y Gómez and among the latter J. Wyatt Durham, an invertebrate paleontologist and oil geologist, recently with a doctorate from the University of California. After some scouting with Durham, Stirt found a fossil bone in a brick pit on the outskirts of Bogotá. It was a *frog* bone, not yet at all what Stirt had gone for! Not long after, in Mosquera, a town on the outskirts of Bogotá, he saw remains of mastodonts not only in a local Salesian school but also in place, associated with fossil horse teeth. These were North American mammals that spread to South America relatively recently. Thus they were not yet what Stirt had hoped to find, although he entered in his journal that "the geology of his area will be discussed after it is thoroughly studied."

On 13 October, more than two months after sailing from San Pedro, Stirt wrote, "At last we left Bogotá for the field." The party included a Colombian contingent, especially Royo y Gómez (generally called Royo in Stirt's journal), Manuel Varón, a cartographer, José E. S. Perico as a field assistant, and Alejandro Comargo as a chauffeur who also hunted fossils. They set out during the rainy season, and tents were not practical in the downpour, so they rented a house in the village of Carmen de Apicalá, southwest of Bogotá and a few miles east of the great Magdalena river. Working out from there Stirt and Royo could examine the thick series of sedimentary rocks that had been named the Honda formation by oil geologists. On 18 October they found a large mammal vertebra, a hopeful sign but not identifiable. Continuing to prospect, they turned up fairly abundant fossil remains of tortoises and (freshwater) crocodiles.

On 30 and 31 October they found some isolated mammal bones and some imperfect teeth. In his journal Stirt compared them with genera from Argentina as old as 20 to 25 million years, but added, "These field identifications are likely to be rather poor." After further study in Berkeley Stirt found this fauna to be of late Miocene age, hence not more than 10 million years old.

In the midst of the rains and the slow going of exploration, Stirt consoled himself with bird watching, the avifauna being of special interest to this former teacher of ornithology. On 9 November they moved southwest from the area around Carmen de Apicalá to Coyaima, on a western tributary of the Magdalena. An oil geologist had told them that there was a veritable "cemetary" (Stirt's spelling) of bones near there. Stirt sadly reported, "The cemetary was an old indian cemetary down by the river." During the next few weeks they found some potentially identifiable mammalian fossils which, later study indicated, could be correlated at least approximately with the Colhuehuapian of Argentina—hence about 23 or 24 million years old. The Coyaima collection thus already indicated that there had been no complete aquatic barrier between northern and southern South America. It also provided evidence that the land connection between North and South America had not yet arisen at that time.

On 28-29 November the party moved some distance westward to the village of Chaparral. Northeast of the village itself near the minor river Tetuán there were old sediments in which some mammalian fossils were found. In the field Stirt thought these might be Eocene in age. On closer study, later, he correlated this fauna with the southern Deseadan land mammal age—hence early Oligocene and about 35 million years old—not quite as old as the Eocene but the oldest mammals then known from northern South America.

In December the party moved to Villavieja (old village) on the east side of the Magdalena river. Stirt had come down with "yellow jaundice" (hepatitis) and on 19 December he went back to Bogotá for treatment and convalescence. In his absence the Colombians explored a large area of badland exposures east and northeast of Villavieja and found numerous and relatively well preserved remains of fossil mammals. These included fragmentary remains of the skeleton of an early monkey found by Manuel I. Varón and José Royo González, Colombian assistants with the party. In 1950 this was named as a new genus and species, *Cebupithecus sarmientoi* by Stirt and Donald E. Savage, who will figure later in this chapter. The species was named for Roberto Sarmiento Soto, then director of the Colombian Servicio Geológico Nacional, which cosponsored all of Stirt's work in that country. In 1951 this specimen was described in more detail by Stirt alone.

Having apparently had a recovery unusually fast for hepatitis, Stirt was back with the field party by the middle of January 1945. They continued successful collecting in the pre-Pleistocene Honda series in the badlands "80 minutes [on muleback] out of Villavieja." Stirt wrote in his journal, "I shall call this the Villavieja fauna." However, he later changed the name to "La Venta fauna," and as such it is now designated worldwide by vertebrate paleontologists. The name is from that of the Quebrada de La Venta. (In Colombia a *quebrada* is a brook, in this case a brook in the badlands tributary to the Río de Las Lajas, which in turn is a tributary of the Magdalena. A *venta* is a small inn or store, equivalent to what is called a *boliche* in Patagonia. *Lajas* are stone slabs. Stirt's journal calls this "Laha," which is not pronounceable in Spanish.

After successful collecting there, Stirt returned to Berkeley. By early summer of the next year, on 18 May 1946, he was back in Bogotá. With a group of oil company geophysicists he went on a reconnaissance trip in a two-motor, five-passenger plane. On 25 May he started for the field by going to "Villavo," as he here designated Villavieja in his journal. Thereafter until 10 August Stirt with some oil company employees and local camp men moved about by planes, trucks, mules, and every other sort of transport. Again he was in Colombia for the long rainy season, and everyone was thoroughly miserable. As a sample, one day's journal entry (for 31 May) is quoted:

> Rained twice during the night—looks like it has set in steady this morning. This camp arrangement with a large tarp spread over poles and an Abercrombie tent pitched below is quite satisfactory but with persistent rains, dampness penetrates. My camp is pitched directly behind the shack called house, here at Bavaria—my crew and provisions

> are in the house or at least out of the rain. Pigs and chickens are all about but they are positively excluded from my Abercrombie although they seek shelter under the tarp. For a certain fee, of course the dueno [*dueño,* "owner"] here has promised to show us where we can cross the river with our provisions and equipment. There are places he says that are shallower than the main crossing. Imagine my surprise yesterday evening to see a Troco truck come pulling in here after we had slopped over the trail for about three hours with mules dropping down or having their packs adjusted frequently . . . my capitas [*capataz,* "boss" of the camp men] Jose [José;] Moreno has gone out with an aareria [*arriero,* "muleteer"] to find the best place for us to cross tomorrow morning.

The "Abercrombie tent" was probably one purchased from the sportsman's store of that name. "Bavaria" clearly was not a village, settlement, or estancia in the usual meaning of that word. It was probably a modest house or hut such as is called a *rancho* in Colombia. "Troco" was the Tropical Oil Company, prominent among the six oil companies that helped to support Stirt's work in Colombia. The river they planned to cross was called by Stirt the Río Guaquive. I have not been able to locate this, but it is probably one of the many tributaries of the Magdalena. Eight of the party went off on muleback the next day, found that the river could be crossed but that the jungle trail was impossible and that there were fourteen more rivers to cross. So they sensibly gave up and retreated.

It would not be fair to say that thereafter the party wandered at random until 10 August. They went where they could. Stirt searched for fossils and made tentative stratigraphic observations, but wrote, "Any discussion of diastrophic events better await a more careful survey and the finding of fossils." He also observed many living birds and mammals and shot and collected some of them, but he wrote that serious zoological observation and collection must await the finding of fossils. Finally in late July and early August they prospected in an oil field at and around El Centro on the upper Magdalena. There Stirt did finally collect some fossils, mostly poorly preserved invertebrates and two turtle bones, one crocodile rib, and numerous (but practically unidentifiable) fish and crocodile teeth. He also found the only fossil mammal of the whole expedition: "What was left of a badly eroded Astrapothere lower jaw" with one broken tooth.

On 10 August 1946 Stirt gave up for the time being. His journal ends on that date with the final statement: "This closes the most expensive and yet most unproductive field season I have ever experienced." In contrast the 1944–45 expedition with its findings had already virtually made the essential points. The fossils collected then had shown that until about eight million years ago the mammals of northern South America were of the same general kinds as those known in southern South America. Therefore the hypothetical strait across the middle of that continent had not existed during most and probably all of the Age of Mammals. It had also been shown to be probable that connection (or reconnection) with North America was within about those last eight million years or, geologically speaking, not much more.

Despite his discouragement in 1946, however, Stirt was stubborn and persistent; he led expeditions back to Colombia twice more, in 1949 and 1951.

In 1949 the work began in late January, and some members carried on until July of that year although Stirt himself left in late February. This expedition was sponsored jointly by the Servicio Geológico Nacional and the University of California Associates in Tropical Biogeography. Diego Henao de Londoño, stratigrapher, represented the Servicio. Perico and González, who had been with the 1944–45 expedition, were also with this one. Stirt's assistant was Robert (Bob) William Fields, a student at the University of California, Berkeley, who went on to get his Ph.D. there in 1952. Charles Mason, a botanist, and Alden Miller, a zoologist, were also with the expedition for some time but were interested in living plants and animals (especially birds) and only incidentally in fossils.

The party left Bogotá by automobile on 26 January and drove to Coyaima, where a fossil mammal of the notoungulate genus *Cochilius* had been found on Stirt's first expedition to Colombia. After a short stop there on 27 January they went on to Chaparral, near which the first expedition had also found fossil mammals. From there to Villavieja they were on the road for two days and nights, sleeping out in hammocks and once being rained on, although for a change Stirt was now in Colombia in the (relatively) dry season. The car they were driving, which had earlier sheared off a wheel, got stuck in fording a stream and had to be pulled out by a tractor. They were charged 30 Colombian pesos for this service, and with water all around Stirt and Bob paid half a peso to get the car's radiator filled.

On 29 January the party reached Villavieja, where some Colombian camp men and the field equipment awaited them. After failing to locate a suitable camping place within walking distance of the La Venta badland, they chose a site near the railroad, on the Quebrada Cerbatana. (This is a brook that empties directly into the Magdalena. A *cerbatena* is an Indian blow-gun.) Stirt wrote, "We shall be short on food until camp can be established." They were saved from starvation by Perico, who brought some "yuca" (cassava) from a passer-by which "he combined with a dove and quail shot by Miller to make us some soup." Stirt added, "Have to rough it one more night." Camp was set up by Henao and other Colombians on 1 February, and with Stirt they "covered the La Venta exposures so I could point out the important localities and . . . especially to show him [Fields] the section as I had divided it [in 1945] to keep the stratigraphic information accurate. . . . Most of the original fossil sites were readily located. We walked about 12 Km. today."

To keep stratigraphic records Stirt had provisionally given names to fossil localities in the long cross-section of the Honda series or formation. The "Monkey unit," between the quebradas Cerbatana and La Venta, was the area and level where the first fossil monkey was found in 1945. Two more were found in that unit in 1948. Another area and level were named for *Miocochilius,* a new genus of notoungulates found there in abundance. At still another, the "toxodont locality," the most important finds were not toxodonts (notoungulates) but rodents of the genus *Scleromys*. This is an example of the increasing evidence of relationships between northern and southern faunas in South America. *Scleromys* had been named on the basis of an Argentine specimen by Florentino Ameghino in 1887. In 1940 Hans Stehlin had described and named a species of the same genus based on a specimen found near Carmen de Apicalá in Colombia and sent to him in Switzerland. Later (in 1957) Fields added to knowledge of Stehlin's species and named another one of that genus from specimens found at Stirton's "Toxodont locality" in the "Monkey unit" of the La Venta badlands.

Stirt had occasion on his Colombian expeditions to complain that assistants and camp men would not work on Sundays. A typical example was 6 February 1949, a Sunday. His journal entry was: "Stayed in camp today. The land owner joined us for lunch and tried to get us drunk on aguadinte but we prevailed." The misspelling of *aguardiente* may suggest that they didn't completely prevail.

Through 21 February 1949 Stirt took Fields and Henao over the whole region in and adjoining the La Venta badlands, going about mostly on horseback. He showed them localities where fossils had been found in 1945, and they found and collected a number of fossil mammals as they went along. The prospect for making a large and important total collection was clearly promising.

Stirt's last days in camp were recorded by him as follows: "These two days were spent helping Alden Miller look for certain birds he was interested in. Then too we always get a great deal of pleasure hunting together so we could hardly deny ourselves this last opportunity." On 24 February Stirt departed for Bogotá and home, leaving Fields and Henao to carry on the field work until July.

Fields, then twenty-eight years old, very ably collected fossils, made a long cross-section of the area, and also, with Henao, prepared a detailed geological map. He published these results of his work in 1959, and already in 1957 he had published a monograph on the fossil rodents collected by the expeditions in Colombia. After getting his doctorate at Berkeley, Fields worked for an oil company for a few years and then joined the faculty of the University of Montana, where as this is written he still is.

In 1950 there was another joint expedition to the upper Magdalena valley by the Colombian Servicio Geológico Nacional and the University of California's Associates in Tropical Biogeography. This time Stirt was not on it. Vertebrate paleontology and the University of California were represented by Donald E. (Don) Savage, who had received his doctorate from that university and was on the staff of its Museum of Paleontology, as he still is. Diego Henao de Londoño again acted on behalf of the Servicio as geologist and manager of the field group. Perico, now an old hand, was with this party. As later reported by Don Savage, the most important paleontological discoveries in the La Venta fauna were good specimens of notoungulates (a toxodont, leontiniids, and an interathere), a large native South American ungulate of a different group (an astrapothere), and a large ground sloth.

In 1951 Stirt made his fourth expedition to Colombia. (He called it his third because he preferred not to count the 1946 failure.) Early in 1951 he went to Bogotá again, and after some scounting around that city he moved on to Mosquera, not far away. (In Colombia there is another Mosquera on the Pacific coast, west of the Andes.) Here, with former companions, Royo, Perico, and Peña, he set up camp on the property of a friendly and helpful Spaniard, Francisco García. In the nearby Cañón de las Cátedras (literally "professorships canyon"—a baffling name), four kilometers southwest of Mosquera, on 23 January they started collecting a fossil fauna consisting largely of mastodonts and horses, which had reached South from North America only about a million years earlier, also glyptodonts and ground sloths, of South American origin but here not long before their extinction.

Some Colombian geologists had considered this formation with its fossils as having been deposited at a lower, savanna level and later having been uplifted as the Andes rose. It is now in the Andes at a height somewhat over 8,600 feet. Stirt found good geological evidence that these sediments had in fact been deposited after the main Andean uplift and at or near their present elevation.

On Sunday, 28 January, after watching bulls from neighboring pastures being sent off for the Sunday bullfights in Bogotá, Stirt and some companions looked over an area where there were many huge rocks. Stirt wrote in his journal:

> Evidently the waters of an ancient Bogotá lake undercut the boulders and in some places tipped them at a marked degree. Some of the early Indians evidently used this as a ceremonial ground[.] [T]here are many red pictographica, more apparent when the surface of the rock is dampened. This area is now set aside as a park and the future Instituto Etnológico will be built here.

During the next few days they explored an old quarry where fossil mammals had been collected. Then on the last day of January Stirt put "the boys" to work "removing the overburden at what we hope will be the mastodont quarry." They got down to the bone level on 3 February, and the first things found in "the mastodont quarry" were parts of a ground sloth. Some fragments of mastodonts did turn up there and elsewhere, but on the whole the collecting was not very good. The party was working so near Bogotá that Stirt and others could occasionally run in there for an afternoon. This was especially handy when Stirt had a fall and was in considerable pain. He could be examined in Bogotá and reassured that none of his bones were broken.

Although Stirt continued until 26 February assiduously collecting a monotonous and rather poorly preserved Pleistocene fauna, he was also collecting Recent mammals and birds for the Museum of Vertebrate Zoology in Berkeley. At least twice during this period he discussed with Dr. "Enriche" (Enrique?) Hubach, then director of the Servicio Geológico Nacional, the possibility of another expedition in Colombia in 1952. (It may here be said at once that this expedition did not occur.) Before leaving Colombia for the last time, Stirt made a quick visit on 28 February and 1 March to Coyaima and Carmen de Apicalá, fossil localities found on his first Colombian expedition.

Stirt's principal interest in the late Pleistocene Mosquera fauna was to find out whether the mammalian faunas then living, as he believed, at levels above 8,600 feet were different from those of about the same age already known at much lower levels in Ecuador and Venezuela. Stirt, and as far as I know anyone else, did not publish on this point. It may be inferred either that the high altitude fauna was not different from that at lower altitudes or that the collection made around Mosquera was inadequate to check that possibility.

The pre-Pleistocene collections made in Colombia under the leadership of Stirt and of Don Savage made highly important additions to knowledge of early mammalian faunas in South America. That is especially true of the great La Venta fauna, although the sparser, even earlier faunas are also significant. Among other publications, the following are particularly important:

Stirton on *Ceboid Monkeys from the Miocene of Colombia* (1951), *A New Genus of Interotheres from the Miocene of Colombia* (1952), and *Vertebrate Paleontology and Continental Stratigraphy in Colombia* (1953).

Fields on *Hystricomorph Rodents from the Late Miocene of Colombia* (1957).

Savage, *Report on Fossil Vertebrates from the Upper Magdalena Valley, Colombia* (1951).

Malcolm C. McKenna, *Survival of Primitive Notoungulates and Condylarths into the Miocene of Colombia* (1956, based on the collections made by Stirton and Savage).

While he was in Bogotá in late March 1951 Stirt had discussed with Roberto Sarmiento Soto "the possibility of [his] going to Australia on [his] first sabbatical leave from the University." Sarmiento Soto suggested that Stirt discuss instead the possibility of another Colombian expedition. As already noted, Stirt did so. Nevertheless Australia won out. The Pleistocene mammalian fauna of Australia had been fairly well known for a century or so, but extremely little was known of pre-Pleistocene mammals on that continent. Stirt conceived the audacious idea—or, as it has been called, "the desperate gamble"—of going to South and Central Australia in search of Tertiary (pre-Pleistocene) mammals. In 1953 he did go, and he did discover pre-Pleistocene mammals. The series of expeditions that he started has continued down to the present time under varying leadership.

On 14 June 1966, while attending a Southern California meeting of the American Society of Mammalogists, Stirt, aged sixty-four, died from a massive heart attack. His wife and son survived him.

In spite of his many foreign trips, including one to Mauritius and South Africa, Stirt continued throughout his life to study North American fossil mammals. At the time of his death there were six research papers of his in press. Five of these were on Australian fossil mammals, and one was on a North American extinct family (Protoceratidae). It should be added that in 1959 his textbook *Time, Life, and Man: The Fossil Record* had been published.

Notes

For the purpose of this chapter William A. Clemens provided xeroxes of the relevant parts of Stirt's four Colombian journals and field notes. I have also read all of his publications, including contributions to the *News Bulletin* of the Society of Vertebrate Paleontology, of which he had been a president. Obituaries by his colleagues at the University of California appeared in that *News Bulletin* and in the *Journal of Mammalogy*. I was also long personally acquainted with Stirt, who was a little less than one year older than I, and I am still acquainted with Don Savage, who led one of the Colombia expeditions.

Paula Couto

On 30 August 1910 Carlos de Paula Couto, son of Tito de Paula Couto and Julieta Silva Couto, was born in Porto Alegre, the pleasant capital of Rio Grande do Sul, the southernmost state of Brazil. His formal education was in the military school of that city (Colégio Militar de Porto Alegre). As early as 1923, when he was still a pupil in that school, he had become fascinated by paleontology, to which he was devoted for the rest of his life. At that time no advanced course in geology and paleontology was being given in Porto Alegre, and young Carlos had to pursue self-education in those fields with what he could find in libraries and see in some local collections.

Carlos (as it seems most natural for me to call him here—we were on first-name terms for many years) married Zilah Barcellos on 2 July 1938. They had two sons, Tito and Claudio. His wife and sons survived him.

During the years 1936–1944 Carlos was employed as bookkeeper of the fiscal commission of the national treasury for Rio Grande do Sul, in other words the office of the federal tax collector for that state. By 1937 he was already a sophisticated paleontologist, at least insofar as vertebrate fossils of Rio Grande do Sul were concerned. In that year he published the first of what would eventually be more than two hundred scientific publications. This one, on fossil ground sloths (*gravígrados*), appeared in the journal of the Instituto Histórico e Geográfico do Rio Grande do Sul in Porto Alegre. His major work during the period was entitled *Paleontologia do Rio Grande do Sul,* with 203 text pages and 26 plates, plus bibliography and two indexes. In spite of the breadth of the title, this work devoted less than one page to fossil invertebrates, more to fossil reptiles, and most to fossil mammals, not restricted to those known from Rio Grande do Sul. This was first published in the journal of the Institute and then as a separate paperback book.

In that same period of his life Carlos suggested to Mathias Gonçalves de Oliveira Roxo that there should be a law controlling the collection of fossils. Gonçalves, as head of the Section of Paleontology in the Division of Geology and Mineralogy of the Ministry of Agriculture, passed this suggestion on to the national government, and it resulted in the following decree by Getulio Vargas, who was then president of Brazil and was ruling by decree:

> Decree—Law No. 4, 146—of 4 March 1942
>
> The President of the Republic, using the power conferred on him by article 180 of the Constitution, decrees:
>
> Article 1. The fossil deposits are properties of the Nation, and, as such, the extraction of fossil specimens depends on previous authorization and superintendence by the National Department of Mineral Production of the Ministry of Agriculture.
>
> Only paragraph. Independent of this authorization and superintendence the explorations of fossil deposits made by national, state and related official establishments, must in this case have previous communication with the National Department of Mineral Production.
>
> Article 2. Provisions contrary to this are revoked.
>
> Rio de Janeiro, 4 March 1942, [year] 121 of Independence and 54 of the Republic

This came to be known as "the Paula Couto Law." Carlos's proposal was intended not so much to keep collection of fossils in the hands of the government as to prevent their collection by inexperienced and often destructive hands. This was embodied in a statement published by him, in part translated here:

> There are no doubts that certain individuals exist who, calling themselves paleontologists, ferret out fossil deposits with the sole purpose of making a deal with what fossils they chance to succeed in collecting. They are cunning fellows whose iniquitous action needs to be restrained as a safeguard of the national scientific inheritance.

Paula Couto himself did not insist on strict application of the "Paula Couto Law" to collection of fossils by skilled paleontologists working for strictly scientific purposes, even if they sold or exported fossils.

In 1944 the Museu Nacional in Rio de Janeiro, then capital of Brazil, had a competition for a place as naturalist in the Division of Geology under the Ministry of Education and Culture. Carlos passed with praise and the highest grade and was appointed to the position. In due course he was to become the head (*Chefe*) of the Division of Geology and Mineralogy in that museum, a post he held from 1960 to 1965. During most of his stay there the Museu Nacional was in the former palace of Dom Pedro Segundo (the second emperor of Brazil, who reigned until 1889, when the country became a republic). This palace, remodeled for use as a museum, is in the Quinta Boa Vista (literally "Pleasant View Farm," in fact a large park).

In Rio during that period there were two governmental units and buidings separately concerned with fossils, the Museu Nacional in the Quinta Boa Vista and the Divisão de Geologia e Mineralogia in the Departamento Nacional de Produçao Mineral, or DNPM for short on the Avenida Pasteur in the Bairro Praia Vermelha ("District Red Beach"—*bairro* is the Portuguese version of what is called a *barrio* in Spanish, and this one, like "Copacabana," is named for the beach that it fronts). The museum and this division of the DNPM are far apart, separated by the downtown commercial section of Rio. The vertebrate paleontologist of the DNPM while Carlos de Paula Couto was at the museum was Llewellyn Ivor Price, generally known as Llew, pronouced "loo," as the Welsh *ll* is practically impossible for anyone not Welsh to pronounce.

Llew Price, like Carlos de Paula Couto, was born in the state of Rio Grande do Sul but in a different city, Santa Maria, famous for the early fossil reptiles found there. Llew was thus a native Brazilian, but his parents were from the United States of America, and he grew up as bilingual, speaking and writing both Brazilian Portuguese and North American English perfectly. Following up what might be called an almost congenital interest in fossil reptiles, he went to the United States and studied vertebrate paleontology at the University of Chicago with Alfred Sherwood Romer, already a leading authority on fossil reptiles. Llew had somehow developed skill in drawing illustrations. As Romer stated, Llew drew "the greater part of the original illustrations" and further did "painstaking work as a collaborator" on the first edition (1933) of Romer's outstanding textbook *Vertebrate Paleontology*. There were two later extensively revised editions (1945 and 1966), and Llew's drawings still appeared in them but were not added to by him, as he had returned to Brazil.

As the years went on, both the Museu Nacional and the DNPM accumulated large collections of fossil vertebrates, including mammals and reptiles. The separation of these collections in two nonadjacent institutions was somewhat inconvenient, but the resident vertebrate paleontologists were friendly, and they reached an agreement that as a rule Carlos would study the fossil mammals in either collection and Llew the fossil reptiles, also in either one. Carlos collected relatively few fossil reptiles for the museum and Llew did collect or acquire some important fossil mammals for the DNPM, but these were generally studied by Carlos or in a few instances, to be mentioned more specifically later, by Carlos and a collaborator. Llew was temperamentally more interested in drawing and in field work than in writing. From 1940 to 1973 he published about twenty-five papers on Brazilian fossil vertebrates, most of them notes or short descriptions of fossil reptiles. He drew a large number of illustrations of Brazilian fossils, almost all of fossil reptiles, but as far as I have been able to learn only a few of these were, or have now been, published.

Carlos was more prolific both in publications of South American fossils per year and in years of publishing on them and some related subjects. My records indicate that his first paper was published in 1937, as mentioned earlier, and his last in 1982, although it is possible that some may still be published posthumously. Almost all his publications were on fossil mammals, sometimes a single specimen or collection of them, sometimes whole faunas or large groups in South America or world-wide, and at last, as will be shown here, on all fossil mammals everywhere.

A project on which Carlos and Llew, Museu Nacional and DNPM, collaborated for years was the vertebrate fauna from Itaboraí. Itaboraí is a village northeast of the city of Niterói, which is on the east side of the harbor of Rio de Janeiro at the entrance to enormous Guanabara Bay. In the general vicinity of Itaboraí, or more exactly on a former farm or ranch (*fazenda* in Portuguese) called São José de Itaboraí, there is a depression about a kilometer in length and half that in width, filled with limestone to a depth of about a hundred meters. Limestone is a necessary ingredient for making cement. There is no other limestone within a great distance from Rio, and once it was determined in the 1920s that this was indeed limestone the Companhia Nacional de Cimento Portland began quarrying it on a large scale. The first fossils found in the limestone were snails, which eventually were sent to a North American invertebrate paleontologist, Carlotta J. Maury. Maury eventually (in 1935) published on these as late Miocene or early Pliocene. As will appear in the next paragraph, this was grossly incorrect: the limestone cannot be later than Paleocene. It may be digressive here, but not malapropos, to note that Maury in the same year (1935) erred in the opposite direction: she misidentified mollusks from the Amazon Basin and considered them some millions of years in age (Pliocene) whereas they were geologically Recent.

The first remains of fossil vertebrates were found in the Itaboraí limestone basin by an engineer, Paulo Rumiantseff, and eventually deposited in the geological and mineralogical division of the DNPM, where Llew and Carlos both studied them. They included both reptiles and mammals, and in 1946 a joint publication by Price and Paula Couto identified some of them and indicated that they were definitely of early Tertiary, probably Eocene, age. Thereafter at various times parties from both the DNPM and the Museu Nacional made large collections of fossil vertebrates in the quarry. From 1946 to 1979 Carlos published thirteen papers on the geology of the deposit and more extensively on its mammalian fauna. In 1958 he could already list twenty-three mammalian species, all named by him and as far as known then (in 1953) all confined to the Itaboraí fauna, although of the twenty-three genera then listed six were known from the Paleocene of Patagonia.

By 1979 one of those had been renamed as a strictly Itaborian genus, but another genus first named with a Patagonian type-species had been added to the Itaboraian list.

Almost all of the identifiable Itaboraí reptilian and mammalian fossils were found in the filling of solution channels and caves in the limestone. This indicates that the limestone could not be younger than the vertebrate fauna in geological age. However, a few fossils of that fauna were also found embedded in the limestone itself, thus suggesting that the limestone might not be of geologically perceptible age older than the fauna. The alternative, which has been suggested but not proved, is the possibility that the particular pieces of limestone with embedded fossil vertebrates were redeposited from the lime-saturated underground water that had dissolved the original limestone in the channels and caves. The great majority of the fossil bones and teeth were embedded in earthy fillings of the cavities in the limestone, and the best preserved fossils were from the relatively higher of such fillings. The richness of the fauna and these interesting stratigraphic problems for a time have made, and perhaps even now do make, this locality a sort of mecca for paleontologists, one which I visited, long ago now, with Carlos and others.

It is of interest here that in the long series of papers by Carlos on this fauna three in 1952 and one in 1954 were written in English and published by the American Museum of Natural History. Carlos spoke English fluently with only a slight and pleasant Brazilian Portuguese accent, and he wrote it somewhat scholastically but better than the average English-speaking North Americans. He also spoke but did not publish in Argentine Spanish. (Brazilian Portuguese and Argentine Spanish differ much more than people who do not speak or read both realize.)

Carlos's final conclusion on the age of the Itaboraian fauna was that it was Paleocene (pre-Eocene Cenozoic) in age or approximately equal to the Riochican (or "Riochiquense") of Patagonia. It now seems probable that the larger fauna from Itaboraí is somewhat earlier than the type Riochican in Patagonia. If this is corroborated, it might be useful to designate the Itaboraian as a distinct South American land mammal age. The term "Itaboraíense," which would be "Itaboraian" in English, was used by Carlos, but he did not designate it as an age distinct from Riochican.

In 1950–51 Carlos was a John Simon Guggenheim Memorial Fellow and spent that time in North America working in various museums but mostly in the American Museum of Natural History in New York. The publications in English on the Itaboraí fauna, mentioned above, were written at that time. The American Museum had a large collection of fossil ground sloths, mostly collected by Barnum Brown in Cuba. William Diller Matthew had studied these before he left the museum in 1927 and had drafted an incomplete monograph about them but had not finished it when he died in 1930.

In 1950, twenty years later, Carlos completed that study but incorporated in his publication Matthew's work, word by word, identified as such. This was finally published by the American Museum in 1959 with Matthew indicated as senior and Paula Couto as junior authors—a strange collaboration by two paleontologists who never met and one of whom had died long before.

Although not literally themselves South American, the West Indian extinct ground sloths were certainly of South American origin. Ground sloths also of South American origin spread to North America, but the West Indian ground sloths, quite distinct from the North American ones, almost certainly reached the West Indies directly from South America. Having worked over the Cuban forms in 1951, Carlos went on to acquire knowledge of all those in the Greater Antilles as a whole, notably Cuba, Hispaniola (Haiti and the Dominican Republic), Puerto Rico, and even Curaçao, which is not considered a part of the Greater Antilles. Carlos wrote up all this in English in 1967, and it, too, was published by the American Museum. He came to the conclusion (as I had in 1956) that the ground sloths must have reached the Antilles overseas by waif or sweepstakes dispersal, probably in the Miocene, and that the islands involved were not at that time connected by land to each other or to either North or South America.

Here I must go back a bit on following the several main tracks in Carlos's years of studies of fossil mammals. As early as 1946 he had become interested in the collections made in Brazil by Lund in the nineteenth century and later studied by Winge in Denmark. As here related in Chapter 3, memoirs and biographical notes relative to this collection had been translated in large part first into French, then into Portuguese, reviewed and commentated by Paula Couto, and finally published in 1950 by the National Institute of the Book under the Brazilian Ministry of Education and Health. Carlos also published on his own several papers concerning Lund and Winge. He made visits to Lagoa Santa and to the *lapas* (caves) in that region, sometimes with Harold V. Walter, the vice-consul of Great Britain in Belo Horizonte. I find it mildly amusing that the eponym of the "Paula Couto Law" cooperated with Walter, a rank amateur in paleontology who was collecting fossils from the lapas and who sold some of them but refused to give others to the national museum.

In 1956 Carlos was one of several members of a joint expedition of the Museu Nacional, the University of Minas Gerais, and, in the person of Wesley R. Hurt, of the University of South Dakota in North America. Their main aim was to follow so many years later in the footsteps of Lund and to investigate the age of what had been called "Lagoa Santa Man." They worked especially in the Caetano lapas or caverns. As the paleontologist of the primarily ethnological and anthropological party, Carlos later (1958) reported that there was no evidence of humans or their activities contemporaneous with any of the extinct species of other animals in the lapas. (In 1961 Hurt published a paper on "The Cultural Complexes from the Lagoa Santa Region, Brazil.")

This side of Paula Couto's interests culminated in 1980 when he went to Denmark as the representative of the National Academy of Sciences of Brazil at the ceremonies on the hundreth anniversary of Lund's death. (Carlos had been a member of the Academy since 1950.) Queen Margarethe II of Denmark took this occasion to bestow on Carlos de Paula Couto the cross of the Royal Order of Danebrog, a very old decoration, "in gratitude for his notable scientific work on the eminent Danish naturalist"—Lund, of course. It was also in 1980 that the University of Minas Gerais gave Carlos an honorary doctorate. An enthusiastic newspaper published this news under the headline "Pesquisador gaúcho ganha título da Universidade-MG," which can be loosely translated as "Researcher from Rio Grande do Sul wins a title from the University of Minas Gerais." The Brazilian-Portuguese word *gaúcho* is pronounced gah-oó-shoo, unlike the Argentine-Spanish pronunciation of *gaucho,* and although in Brazil it does mean "cowboy" it is also the nickname for anyone from Rio Grande do Sul even if he never saw a cow. "MG" is of course short for "Minas Gerais."

Now back again first to mention two important and lengthy publications by Paula Couto: in 1952 on "The Successive Faunas of Terrestrial Mammals in the American Continent" ("the American Continent" here means both North and South America), and "Brazilian Paleontology, Mammals." The former was one of the "Separate Publications" of the Museu Nacional, and the latter a paperbound book in the "Brazilian Scientific Library" published under the same auspices as the big book mostly about Lund.

As early as 1944 Carlos had published briefly on a fossil mammal from the Amazon basin. In 1956 he published more fully on that subject in a 121-page bulletin of the Brazilian National Research Council (Conselho Nacional de Pesquisas). Here he reviewed all the known fossil mammals from the Cenozoic of "Amazonia" (the Amazon Basin). The specimens were almost all from the upper Rio Juruá and the upper Rio Purus, both tributaries to the Amazon from the Southwest. Those studied or reviewed were in the two fossil vertebrate collections in Rio, the American Museum of Natural History in New York, and the Museu Paraense Emilio Goeldi in Belém do Pará. They were all incomplete, but Carlos was able to identify eight species (one doubtful) of different genera and three genera of specimens not identifiable to species. He indicated that some were Pliocene but most were younger, in the Pleistocene.

In that same year, 1956, an expedition went to the upper Rio Juruá with four non-Amazonian members devoted to search for fossil mammals, to map their occurrences, and to make stratigraphic cross-sections of the river's banks, as exposed at low water. (In the rainy season the banks are entirely under water.) Llew Price represented the DNPM, and the American Museum of Natural History sent me and my field and laboratory assistant George Whitaker. (A graduate student of Columbia University also went with us.) We had partial support from the National Research Council of the U.S.A. and from the National Research Institute of Amazonia (Instituto Nacional de Pesquisas da Amazonia). We collected a fairly large number of fossil mammal specimens, mostly fragmentary but identifiable, and also fewer but better fossil reptiles, skulls of crocodilians and shells of turtles. The fossil reptiles were sent to Rio for Llew to study, and in 1964 he published a paper on the large skull of one of the crocodiles.

Most of the fossil mammals were sent to New York for study but I was unable to work on them for some time, having been seriously injured toward the end of the upper Juruá expedition. I did write some manuscript before I left the American Museum in 1959, but I then arranged for the whole fossil mammal collection and my field and laboratory notes to be sent to Carlos de Paula Couto for further study and publication. Llew Price had gone over the same stretch of the upper Rio Juruá in 1957, and the fossil mammals he then found were also turned over to Carlos for his study. Other researches had priority for some years; then in 1976 publication on the Juruá collections by Carlos de Paula Couto began, in English, but in Brazil. The series of papers was entitled "Fossil Mammals from the Cenozoic of Acre" (Acre is the state where the upper Juruá is). Part I, on the Astrapotheria, an order of large hoofed mammals confined to South America, was read in 1975 at the XXVIII Brazilian Congress of Geology and published the next year in the Annals of the Congress. Part II, on some native South American rodents, was published in *Iheringia,* a Brazilian geological journal, in 1978. Part III, published in 1981, was also in *Iheringia*. It was published as if by Carlos and me as co-authors, but Carlos included what I had written about a large ground sloth (*Eremotherium*) and mastodonts (*Haplomastodon*) more than twenty years earlier. He marked the text actually by me and that by him, the latter on ground sloths, a manatee, a vicugna (*vicuña* in Spanish and English), a peccary, and a tapir. The stratigraphic sections and descriptions and the maps of the upper Juruá are from my field notebook. (Carlos was never personally in that region.) Part IV on some members of the South American and, briefly speaking, Central American order of notoungulates was read at the II Latin-American Paleontological Congress in Porto Alegre in 1981 and published in its Annals. Finally Part V, on other notoungulates and the strictly South American ungulate orders Litopterna, Pyrotheria, and Astrapotheria, was published in *Iheringia* in 1982.

This chapter must now backtrack again to follow another direction of Carlos de Paula Couto's busy and multiple professional life. The occurrence of mastodonts in Brazil has been known since at least 1828, and their presence in other South American countries was known even before that date, as was made evident in the early chapters of the present book. These had been repeatedly studied by leading paleontologists, among others Cabrera (1929) in Argentina, the French vertebrate paleontologist Hoffstetter in Ecuador (especially 1950 and 1952), and, mostly from the literature, by the North American Osborn (especially the first, 1936, volume of his posthumously published massive work on Proboscidea). The problem of the classification of these mastodonts had become extremely complex not only because those and many other paleontologists had recklessly based names of supposedly new genera and species on inadequate specimens but also and especially because the extent of variation in local populations had not been realistically established.

A possibility to make a step toward solution of that main problem arose, unexpectedly, when the Brazilian government undertook the building of a large hotel and spa at a place known as Águas de Araxá. (This at first included a gambling casino, which was later forbidden by Brazilian law.) It was necessary to do some excavation in an ancient stream bed. In April 1944 workmen encountered a pothole in which was a mass of fossil remains of mammals, mostly mastodonts. The Division of Geology and Mineralogy of the DNPM was informed of this discovery and Llewellyn Ivor Price and Rubens da Silva Santos (who does not otherwise figure in the present account) were sent to check on the find. Llew verified the find as including Pleistocene remains of a large number of mastodonts and a few other animals. Some specimens were left not exactly in place as found but installed in a separate small building erected by the hotel builders as a tourist attraction. This was inaugurated in the same year when visited by the president of Brazil and the governor of the State of Minas Geraes.

The even more numerous fossils not included in the exhibit at Águas de Araxá were taken to Rio de Janeiro and stored there in the building of the Division of Geology and Mineralogy. Llew announced the discovery to the Brazilian National Academy of Sciences, and a brief account by him was published in its Annals in September 1944, with the promise that a more detailed account of the site and fauna was in preparation and would be published in due course. However, nothing further was published until 1955. According to the informal agreement between Llew and Carlos, these as fossil mammals although in possession of the DNPM were to be studied by Carlos, whose time was deeply involved in other research for the next ten years or so. At the suggestion of Carlos, the Brazilian National Research Council invited me to spend a few months in Brazil, which I did from October 1954 to January 1955. Carlos had made arrangements for us to work together on the mastodont specimens brought to Rio by Llew.

This collection included teeth from at least thirty individuals, almost certainly a single interbreeding population of the same geological age and living in the same area. We visited the Águas de Araxá and could include in our study the specimens left on exhibition there. With statistical data on variation of one population, and thus one species, we were able to extend identification to other mastodonts, which first we did for all mastodonts known from Brazil. We went together to Belo Horizonte, where there were mastodont specimens from the lapas, some at the university, bought from the British vice-consul Walter, and also some retained by Walter, who permitted us to study them. (This visit was also mentioned in Chapter 3 because of its connection with Lund.) There were a number of specimens from Brazil but not from Águas de Araxá in the Museu Nacional, and of course we included these in our study. Later I studied specimens in the University of São Paulo and in the Instituto Geográfico e Geológico of that city and also in the Museu Júlio de Castilhos in Porto Alegre. From this basis we extended our study to all South American mastodonts, and in 1955 I further extended the study to the Argentine specimens in the museums of Buenos Aires and La Plata. Carlos and I published a Portuguese and English summary of our results in 1955 in the bulletin of the Brazilian National Research Council, and in 1957 our whole joint study in English was published in the bulletin of the American Museum of Natural History.

The reader will have noted that quite varied lines of research had been followed by Carlos de Paulo Couto, and that different lines sometimes tended to overlap in time. It should therefore be mentioned here that in 1972 Carlos moved his residence back to Porto Alegre, without breaking off his connections with Rio de Janeiro. Thus, for example, his research on the Cenozoic fossil mammals of the Amazon Basin was started in Rio but was ultimately completed in Porto Alegre. Now with the end of his life approaching in the city of his birth, we turn to the capstone or masterpiece of that life.

What may justly be considered the climax of Carlos de Paula Couto's industrious and useful life was his *Tratado de Paleomastozoologia* (Treatise of Mammalian Paleozoology) completed in 1978 and published by the Brazilian National Academy of Sciences in 1979. This is a massive quarto volume of 590 pages with 569 illustrations. The introduction starts as follows:

> The book here presented is the first Treatise of Mammalian Paleozoology to be published in Portuguese and to be brought out in Latin America. It is a general synthesis of present knowledge of the paleontology of the mammals treated in the scope of modern systematics.

That is a modest understatement. This was, and still is, the most nearly up-to-date and detailed treatment of the subject in any language. The most nearly comparable publications on the same subject are the two volumes in German on the phylogeny of mammals by Erich Thenius published in 1969 as parts of the *Handbuch der Zoologie* and the last three volumes in

French of the *Traité de Zoologie* written by numerous authors under the editorship of Jean Piveteau and published in 1957 to 1961. There is nothing nearly so comprehensive, detailed, and recent in English, a language that all paleomammalogists can read and in which a plurality of them write. It is astonishing that the most nearly complete and also most nearly current work on fossil mammals should be in a language which relatively few paleomammalogists, even in Latin America, can easily read. Still more astonishing is the fact that it was written by a single author and so near to what proved to be the closing years of his active life.

Carlos de Paula Couto's life and works were highly valued in Brazil and in other countries. To his distinctions already mentioned here it may be added that he held professorships in both the Federal Universities of Rio de Janeiro and that of Rio Grande do Sul, that he was for some years president of the Zoobotanical Foundation of Rio Grande do Sul, and that he received the prestigious José Bonifácio de Andrade e Silva Prize of the Brazilian Geological Society and the Paulo Ericksen de Oliveira Prize of the Brazilian Academy of Sciences.

Carlos was a rather quiet, serious, and even reserved man, in Portuguese terms *reticente,* sometimes *retraído,* and rarely *chistoso,* a term that means "joking" or "facetious" in both Portuguese and Spanish although differently pronounced in those languages. In some contrast with his colleague Llew Price, he generally took most things soberly and seriously. He was a handsome man, and even as he approached old age his hair had retreated only a little and was still dark. He was always clean-shaven, unlike his father who had a magnificent black mustache in the photographs of him that I have seen. Carlos usually wore semiformal or what we speak of as "business" clothes. Although in one sense married to his science, he was also a family man, loving to wife and sons.

After a long, painful, uncomplaining illness Carlos de Paula Couto died of cancer on 15 December 1982.

Notes

As has been made evident, I knew Carlos for many years, spent considerable time with him both in the United States of (North) America and in (the United States of) Brazil. We also corresponded up to a short time before his death. I read all his publications, usually as soon as they were distributed. I had an unpublished curriculum vitae of him in connection with his Guggenheim Fellowship, another written more recently, and a *vida* and bibliography published in 1981 by *Iheringia* in commemoration of his seventieth birthday (in 1980). His wife and now widow Zilah also provided me with information on certain details.

Unfinished Work

The preceding chapters have all centered primarily on paleontologists whose work is now over. Cuvier, Lund, and Darwin died before anyone now living was born. Winge, Gaudry, Florentino Ameghino, and Hatcher lived on into the early twentieth century but died when paleontologists now active were children. I was personally and professionally acquainted with the others especially featured in previous chapters: Scott, Carlos Ameghino, Riggs, Cabrera, Sinclair, Kraglievich, Stirton, Patterson, and Paula Couto (who died after I started to write this book).

The collection and study of South American mammalian fossils has never been so extensive as it is now, with South American, North American, and European paleontologists and stratigraphers involved. They are so many and their work is so varied and extensive that in a single chapter I cannot name them all or adequately review what they are all doing; I can only offer a summary. The subject is also moving so rapidly that what I write now will necessarily be somewhat out of date by the time it is in print.

Dealing with recent and current collections and studies, it is not practical to continue the previous historical and biographical sequence of selected persons and events. This chapter might be organized by institutions involved, but two or more may be cooperating. Organization could also be by geographic regions or geological ages in which the fossils occurred, but again these may be multiple. The approach cannot here be consistently uniform; it must involve something of each of the different possibilities.

First an institution: the Museo de La Plata. In the collection and study of fossil mammals this is at present the most extensive and active South American institution. That is in considerable part because of the close association of the museum with the Universidad Nacional de La Plata. This is decidedly advantageous for both institutions and for the science of paleomammalogy. The museum provides teaching and practical experience for advanced students in the university, and at the same time the students aid in the collection, study, and publication of the museum's present and constantly expanding collections of fossil mammals.

Within the museum the División Paleontología Vertebrados is headed by Rosendo Pascual, an eminent paleomammalogist, skillfully seconded by María Guiomar Vucetich. Those working with them recently include, among others, Pedro Bondesio and Jorge Fernández, especially with Pascual, and Mariano Bond, especially with Vucetich. Juan C. Quiroga should also be mentioned for his continuing research on the special science of paleoneurology, the study of casts of the brain cavities of fossil skulls. He has followed this from mammal-like reptiles to some of the peculiar South American mammals themselves. G. J. Scillato Yané also studies and teaches here. An example of how such teaching in the museum is spreading can be seen in the affiliation of Silvia Aramayo, a student of Scillato Yané's with the Universidad Nacional del Sur in Bahía Blanca.

This museum continues interest in the classic fossil faunas of Patagonia and the generally somewhat later ones elsewhere in Argentina, such as those collected by Riggs and by Cabrera (see Chapters 9 and 10). Some further collections and studies have been made on faunas of Huayquerian age (late Miocene, some five to nine million years old) in the province of La Pampa, which is immediately north of Patagonia, usually so designated, and west of the province of Buenos Aires.

However, the most interesting and novel faunas found and studied by this group in recent years have been in the two farthest northwest provinces: Jujuy (pronounced more or less hoo-hwee, with a coarse *h* rather like a Scotch *ch*) and Salta (pronounced as it looks but with the *a*'s near those usual in the English world "alms"). Jujuy is a relatively small province forming the extreme northwest corner of Argentina, and Salta is a larger, irregularly U-shaped province which looks on the map as if it were clasping or nearly surrounding Jujuy. In that region collectors from La Plata by recent accounts had located three especially rich deposits with fossil mammals somewhat crushed but otherwise exceptionally well preserved. Two of the localities so far well explored are in Salta and one in Jujuy.

These fossils are from varied sedimentary rocks known collectively as the Salta group. In this region the upper part of this group has been divided stratigraphically into the succession Mealla, Maíz Gordo, and Lumbrera formations, in sequence from lowest (hence oldest) to highest (youngest). At last reports no mammals had been found in the Maíz Gordo formation, although it is rich in other vertebrates, mostly turtles. There are also aquatic vertebrates in the Mealla and Lumbrera formations, but the mammals there are evidently not aquatic. They are land mammals that were buried in silt-laden floodwaters.

As the study of these mammals proceeds it is adding notably to knowledge of regional faunas in the early Cenozoic (Age of Mammals) in South America. As so far studied and published, it appears that the fauna in the Mealla formation is approximately of Paleocene age, tentatively referable to the Riochican land mammal age previously known only from Central Patagonia in Argentina and from Itaboraí in Brazil (see Chapter 12). The fauna from the somewhat later Lumbrera formation is similarly correlated with the Casamayoran land mammal age, the earliest part of the Eocene epoch, hitherto known almost entirely from Argentina (Chapters 4–6).

In addition to mammals previously known from the Riochican and the Casamayoran, thus permitting correlation of these beds with the others of those ages, the Mealla and Lumbrera faunas also include a number of species, several genera, and possibly some families not previously known. This indicates that in the late Paleocene and early Eocene there was some regional differentiation of mammalian faunas in South America. The environmental and ecological conditions must have been somewhat different in the region that is now Patagonia and in that now Jujuy and Salta. However, it must be noted that the present ecological differences between these two areas have been radically changed since the early Cenozoic by the great uplift of the Andes, an uplift that had scarcely or not at all started in the early Eocene.

The region that is now Andean, reaching the highest point (Aconcagua, 22,834 feet high) in the Western Hemisphere, also includes the altiplano or high plateau area in northwestern Argentina and western Bolivia. Titicaca, the highest navigable lake in the world, is on the altiplano between Bolivia and Peru at an elevation of 12,506 feet. The regions in Jujuy and Salta where the Museo de La Plata is continuing exploration and collection of early Cenozoic mammals are on the altiplano at elevations between approximately 10,000 and 13,000 feet. Before the Andean uplift, and hence early in the Cenozoic and deployment of varied mammalian faunas, the region now altiplano was near sea level, a shallow nearly south-southwest to north-northeast trough into which streams were periodically depositing silt and sand and incidentally here and there the remains of what are now vestiges of many ancient mammals. Occasionally, too, at least the southern part of the trough included a shallow arm of the sea. These facts will become still more important when later in this chapter we consider discoveries of early mammals in what is now Bolivia, an entirely political separation of continuous geological conditions.

Before leaving discussion of Argentina, three other institutions there especially important for paleontology must be briefly noted. First is the Museo Nacional de Ciencias Naturales in Buenos Aires, frequently mentioned in previous chapters. As regards fossil mammals this is and will always be a mecca for paleontologists, and specifically for paleomammalogists, if only because it houses most of the Ameghino collection, including almost all of Don Florentino's type specimens. The present chief of the section of vertebrate paleontology is José Bonaparte. His own interests have been mainly in Mesozoic reptiles, especially the mammal-like reptiles of the early Mesozoic (Triassic) in Argentina and also Brazil, and lately more particularly Argentine dinosaurs. Under his direction a tremendous skeleton of one of the great Argentine sauropods was recently mounted and installed as the stunning centerpiece of the large and attractive exhibition hall of paleontology.

Work on fossil mammals at this museum has recently been carried on especially by Miguel Hernando Soria (h). He has been working with some of the Patagonian mammals from the Miocene Santa Cruz formation, but also with probably Paleocene mammals from the Río Loro formation in the northwestern province of Tucumán. Before his appointment in Buenos Aires Bonaparte worked for some years in the Instituto Miguel Lillo in the city of Tucumán, capital of the province of that name. Bonaparte still maintains contact with that important institution for research in the various aspects of natural history, and Jaime Eduardo Powell has maintained work there on dinosaurs, in connection with Bonaparte, and also on the Río Loro mammals in connection with Soria.

There are numerous relatively small collections in Argentine provincial universities, several local museums, and a few private collections. For the present subject the most important of these is the Museo Municipal de Ciencias Naturales Lorenzo Scaglia. This is in Mar del Plata, a coastal city near the southeastern point of the province of Buenos Aires. Its name has no direct connection with that of the city of La Plata or the La Plata River (or estuary). It has a resident population of several hundred thousand and is the commercial center for the surrounding pampa and the base of extensive harvesting of fishes and other marine edibles. It is, however, most widely known as one of the most popular seaside resorts in South America. In the summer season the population swells enormously as the city fills with vacationing families from Buenos Aires and widely elsewhere.

For present purposes, however, interest in Mar del Plata is not in its series of fine beaches or its delicious and varied sea food. The basis of the present interest starts at the many miles of steep scarps that rise from the tops of the beaches up to the higher, more level land of pampa, estancias, and the major city Mar del Plata. Those scarps, slowly but constantly being eroded back by waves at high tides, are cross-sections of a series of sediments tens and hundreds of thousands of years old. These contain large numbers of fossils, most of them ancient, now extinct, land mammals. The total stratigraphic sequence here exposed along the present coast runs from late Pliocene through almost the whole of the following Pleistocene. For a paleontologist this is almost or quite the best place in South America to study the faunal sequence during a highly dramatic episode: The Great Interchange of earlier unmixed North and South American mammals. For millions of years the two continents had been separated by an oceanic barrier, and their faunas had evolved differently in the two continents. When they were united by a land route, a faunal interchange, called the Great Interchange by American paleontologists, took place. This was at its height when the sediments of the scarps at and around Mar del Plata were being deposited and burying a splendid record of the fauna this far south as it became altered by the interchange.

A citizen of Mar del Plata, Lorenzo Scaglia, of Italian descent like so many Argentinians, became fascinated by the fossils, especially the mammals, weathering out of the scarps. He made a large, and constantly growing collection of them, which he donated to the city. The collection, open to the public, was for some time rather simply displayed in one of the municipal buildings. Eventually a building exclusively for museum purposes was erected on its own grounds near the shore, with exhibition halls for the public, laboratory place for the preparation and study of fossils, and accessible storage for the many specimens not on exhibition.

Don Lorenzo's son Galileo Scaglia was made, and still is, the director of this fine provincial museum, the name of which was eventually changed to Museo Municipal de Ciencias Naturales Lorenzo Scaglia. As director, Don Galileo makes the remarkable and still growing riches of this museum fully available for study by any competent paleontologist, either at the museum itself or on temporary loan even outside of Argentina. Many paleontologists, including me, have profited by this useful management of the collections, and numerous research papers involving the museum's collections have been published by that museum itself and elsewhere.

In this connection I must mention two Argentine paleontologists who for some time worked for and with this museum, although neither of them has now been there for a number of years. One was the son of Lucas Kraglievich (see Chapter 9 of this book), known both as Jorge Lucas Kraglievich and Lucas Jorge Kraglievich. He made some short studies on fossil mammals and started one more extensive on the extinct, large, flightless, carnivorous birds known as phororhacoids. This project was taken over by Bryan Patterson and published as by Patterson and Kraglievich by the Mar del Plata museum. In the present connection Jorge Kraglievich's most important work, also published by that museum, was a stratigraphic study of the sequence of beds along the coast of Mar del Plata, with special attention to the Chapadmalal formation. This is extensively exposed in this region, where it is the type for the land mammal age Chapadmalalan. In or about 1960 Jorge Kraglievich left Argentina, and I do not know anything about his later whereabouts or activities beyond a bibliographic reference to a paper on some fossil rodents published in France and with a French title in 1965.

Another paleontologist closely associated for a time with the Mar del Plata museum is Osvaldo Reig. Here the most relevant of his studies at that museum was the review and updating of the Chapadmalalan mammalian fauna. He also started more detailed studies of fossil mammals in the Mar del Plata museum, but some of these were passed on to others (including me) for completion. He moved for a time to Chile and later also worked in the United States (of North America) and in England, then settled for some years in Venezuela. His interests have been wide-ranging and their results prolific. His long bibliography includes publications on amphibians, both fossil and living, on various groups of reptiles, mostly fossil, and especially on mammals, earlier mainly fossil but in recent years more on living South American mammals. After settling in Venezuela his main concern came to be studies of karyotypes, the numbers and characteristics of chromosomes, necessarily of recent species, as chromosomes are almost never observable in fossils. He has studied in this way many South American rodents and marsupials, and has undertaken to synthesize the karyotypic evidence with that of fossil tooth and bone morphology. Much of this work has been published, but even more voluminous studies are not yet in print as I write this. His latest approaches to the relationships and classification of related fossil and recent mammals are ingenious and interesting. The results are not yet entirely clear, and judgment as to the eventual outcome must await further publication and discussion. As I write this, Reig has moved back to Argentina after years in Venezuela, and his present affiliation and plans are not known to me. As far as I know he did not collect fossil mammals in Venezuela, and while there his involvement with fossils was mainly based on his own and other publications on fossils found elsewhere.

There are a number of other Argentine students and practitioners of paleomammalogy, but before moving on to other areas and approaches, I shall here mention just one more: Rodolfo M. Casamiquela. Like Reig he somewhat opportunistically started his career with varied studies of fossil frogs and reptiles and has continued to broaden his interests since then. In 1961 he described some fossil footprints from the Jurassic (middle Mesozoic) of Patagonia. These were probably, although not quite certainly, made by small mammals. If so they are by far the oldest known from South America. As mammals equal in age or even older are known from dentition and bones in North America, Europe, and Africa, it is quite likely that they did also occur in South America, but they cannot be clearly identified, correctly named, or classified from the footprints alone. (I shall later enumerate the scanty, much later, classifiable Mesozoic mammals that are now known from South America.) Stimulated by this discovery, Casamiquela has continued to study whatever fossil footprints have come to his attention. In 1974 he published *Estudios Icnológicos* (Studies of the science of fossil footprints), which is practically a textbook—the only one, so far as I know—on this interesting and rather exotic subject, which has a large descriptive literature. Among the most interesting of Casamiquela's ichnological studies is evidence that *Megatherium,* the giant ground sloth usually reconstructed as walking on all four feet, sometimes if not customarily was bipedal, walking on the hind feet only.

During a period of political unrest, Casamiquela, like Reig, emigrated to Chile. While there, and later on observations made there, he published several short works on Chilean fossil mammals. He returned to Argentina and became a resident of Viedma, the capital of the province of Río Negro in northern Patagonia. In addition to his paleontological studies, he has become an authority on the anthropology, archaeology, and linguistics of the native Patagonian Indians.

Up to this point discussion has mainly centered on the discoveries and studies of the fossil mammals of Argentina. The first fossil mammals to become subjects of truly scientific attention were from Argentina (see especially Chapters 1 and 2). Thereafter the building up of the most nearly complete sequence of South American mammalian faunas and their codification in the system of Cenozoic land mammal ages has been, and is still today, based on Argentina. Important finds and studies shorter in geological scope but within that scope of great paleontological and historical interest in other South American countries have also been covered to some extent in earlier chapters (Chapters 3 and 12 for Brazil, 11 for Colombia, and some briefer mentions of other countries in other chapters). Recently, however, the most extensive and interesting work outside of Argentina has been in Bolivia. We now turn more specifically to what has been, and is, going on in that country as regards fossil mammals.

It has been noted that the presence of abundant Pleistocene mammals in the area around Tarija in southern Bolivia was already known in early days of European exploration. Chapter 10, above, was in part devoted to the acquisition of a large collection there for the North American Field Museum of Natural History, and other collections, mostly amateur or commercial, were more briefly mentioned there. The subsequent great extension of knowledge of Bolivian fossil mammals, both geological and geographical, has been principally stimulated and also in considerable part carried out by Robert Hoffstetter in collaboration with some other French paleontologists and with Bolivian stratigraphers and paleontologists, especially those of the Servicio Geológico de Bolivia, generally known as GEOBOL. Connected for a time with GEOBOL was a capable and promising young paleomammalogist, Carlos Villarroel A., who collaborated to some extent with Hoffstetter and later with a largely North American expedition, as well as doing important collecting and studies on his own. It is regrettable that quite recently he found the situation in Bolivia so disaccordant with his aims that he emigrated to Colombia.

In the southern part of the eastern cordilleran region of Bolivia, Hoffstetter added two diverse faunas to the longer and better known one of Tarija. Northeast of Tarija is Ñuapua and southwest, close to the Argentina border, is Quebrada Honda. In 1981 a joint expedition of the Florida State Museum, the Los Angeles County Museum, and GEOBOL revisited those three regions and made good collections of fossil mammals from them.

They also made stratigraphic sections and collected materials for subsequent paleomagnetic and radiometric dating. The results, now published in part, agree in general with Hoffstetter's original dating but are more precise. The main fauna at Ñuapua has a minimum radiocarbon dating of 6600 ± 370 years before present. Stratigraphically it seems to be early Recent rather than late Pleistocene. The fauna includes human (Paleoindian) fossils associated with a mixture of extinct and extant other mammals, some of South and others of North American origin. Turning to the Tarija fauna, it had all generally been considered as of or about the same geological age. Preliminary findings on the new data, however, indicate that it covers a span of approximately 1,000,000 to 600,000 years before present. Hence it is more or less middle Pleistocene, but with a range in time that could well include some faunal change. The principals on this expedition were Bruce J. MacFadden and Ronald G. Wolfe, of the Florida State Museum and of the University of Florida, Kenneth E. Campbell, Jr., of the Los Angeles County Museum of Natural History, and Oscar Siles, a geological engineer from GEOBOL.

From Tarija this party went on to Quebrada Honda, where they made another large collection of fossil mammals and also took some rock samples for paleomagnetic determinations. On the basis of the fossils alone, Hoffstetter considered this fauna to be middle Miocene in age, approximating the Friasian land mammal age in Argentina. As I write this, the paleomagnetic evidence has not been published in detail, but the later studies apparently support Hoffstetter's correlation. That makes the age somewhere around 15 million years.

Hoffstetter has also put on the map no fewer than twelve other Bolivian localities with fossil mammals of late Miocene age, approximately Chasicoan and Huayquerian in terms of the land mammal ages established in Argentina. One (Rio Yapacani) is in the Subandine zone and one (Suipacha) is in the southern part of the central cordillera. The others are all quite local and scattered along much of the altiplano, the very high plateau region of Bolivia that borders against Chile along the western cordillera for most of its length and for a shorter distance against Peru in the northwest. Titicaca is there, and the Bolivian-Chilean border runs somewhat irregularly across the middle of that large, high lake. The most southern Bolivian late Miocene locality on the altiplano is between Cerdas and Atocha, and there is also middle Miocene in that area.

Most of the recorded late Miocene localities are somewhat concentrated in a considerable area on the northern part of the Bolivian altiplano, some distance south or more southwest of La Paz, which is the actual, working capital of Bolivia although for odd historical reasons Sucre is the nominal or official capital. The large area involved in this cluster of late Miocene localities can be simply located as more or less between Callapa to the southwest and Achiri to the northwest—localities otherwise so insignificant that even the large, detailed atlases ignore them. Although the mammals from these scattered localities indicate correlation with the Chasicoan and Huayquerian of Argentina, they have not yet been described in detail and the collections apparently have not been large.

A sequence of fossil mammal localities somewhat later in age is known from near La Paz and in a sequence continuing considerably south of there on the Bolivian altiplano. These were collected and studied at first especially by Villarroel and eventually in collaboration with Hoffstetter. They extend as far from La Paz as Umala, a town significant enough to be on most maps of Bolivia. A few miles north of Umala, and hence that much nearer La Paz, is Viscachani, a name more obscure to cartographers. Its greatest interest so far has been the discovery there by Villarroel and subsequent identification and naming by him with Hoffstetter of a strange marsupial closely related to a group otherwise known only from Argentina. The first fragmentary specimens of this group were found by Carlos Ameghino and incorrectly identified by Florentino Ameghino. Much later Galileo Scaglia found near Mar del Plata, Argentina, much of a whole skeleton which he placed in my hands on loan from the museum of which he is director (see above). The animal turned out to represent a distinct family of probable marsupials, remarkably similar in superficial appearance and function to the so-called kangaroo rats. The latter animals are rodents that evolved in North America and are neither kangaroos nor rats despite their vernacular name. They belong to the rodent family Heteromyidae. Their resemblance to the South American marsupial family, Argyrolagidae, is a remarkable instance of evolutionary convergence: animals in different regions and quite different in ancestry sometimes become similar in superficial appearance and in habits when they adapt to similar ecologies and environments.

The presence of this argyrolagid in Bolivia and the ensemble of faunas between Umala and La Paz indicate correlation with the Pliocene of Argentina, including the Montehermosan and Chapadmalalan land mammal ages. Thus by putting together the sequence of widespread local fossil mammalian faunas in Bolivia they provide an almost continuous sequence from middle Miocene to Recent, approximately the last 15 million years or in terms of land mammal ages from Friasian through Lujanian. As will next be noted, mammals much older than middle Miocene or Friasian are now known from Bolivia, but two large gaps in the geological sequence still remain to be filled in.

The next oldest fauna to be noted occurs at Lacayani, southeast of La Paz, and in the Luribay-Salla stratigraphic basin still farther southeast. (All three names are now well known to paleontologists, but I find only Luribay in the big London Times atlas.) The occurrence of mammalian fossils and their approximate age have been known since the 1960s, largely by discoveries by GEOBOL and other Bolivian geologists. Most of the identifications have been made by Robert Hoffstetter, and he has published some details about parts of these faunas, as well as some shorter and less detailed notes. A needed monograph, including all the species present in the various collections, has not yet been published.

The Lacayani and the Luribay-Salla faunas are believed to be of the same geological age, early Oligocene, and to belong in the land mammal age Deseadan. Hence they are about 35 million years old, and about 20 million years older than the middle Miocene Bolivian faunas previously briefly noted here. Despite their near equivalence in age, the Lacayani and the Luribay-Salla faunas are found by Hoffstetter to differ markedly in some respects, especially as regards the rodents in them. He believes the difference to be due to facies or ecological factors rather than to age.

Rodents are first known in South America from the Deseadan in both Argentina and Bolivia, but in both countries (and throughout South America) there is a hiatus of record for the late Eocene. As the rodents were already somewhat differentiated in the Deseadan, it is probable that they somehow reached South America during that hiatus of presently known record. Undoubtedly the most interesting single find in the Luribay-Salla fauna was part of the upper jaw of a monkey with three well preserved teeth, another broken, and the root of a fifth. This was and still is the oldest primate (in this case a monkey) known in South America. That first specimen was found by L. Braniša of the University of La Paz, but he turned it over to Hoffstetter for study and for deposit in the Institute of Paleontology of the Paris National Museum of Natural History, with which as well as the University of Paris Hoffstetter was connected. To honor the discoverer, Hoffstetter named the new genus *Branisella*.

This discovery makes it possible, or probable, that rodents and primates reached South America at about the same time, most likely late Eocene. Hoffstetter has vehemently maintained that both these groups (orders in the Linnaean hierarchy) came to South America from Africa. Other paleontologists have held that one or probably both came from North America. Either way it is a problem, although not an insuperable one, that in the late Eocene there was almost certainly no land connection of South America with either Africa or North America. As we hopefully wait for more and better evidence, this remains (for me, at least) an unsolved problem.

Earlier in this chapter I noted that some Jurassic footprints in Argentina might have been made by mammals but could not be definitely identified on this evidence alone. For a long time the oldest clearly identified and dated mammals from South America were Paleocene in geological age and from several localities in Argentina and one in Brazil. For years paleontologists scanned the Mesozoic, and especially the late Cretaceous, rocks of

South America in hopes of finding mammals. The beginning of realization of these hopes has now (by mid-1983) occurred at three different localities and in three different countries.

The first definitely identifiable Mesozoic mammal in South America was found in 1965 by Maurice Mattauere, accompanying an expedition with other paleontologist-geologists from Montpellier, France, and some Peruvian geologists. This was a fragment of a lower jaw with two broken molar teeth. It was studied in Montpellier by Louis Thaler, a paleomammalogist, who gave it the name *Perutherium altiplanense,* "the Peruvian mammal from the altiplano." From the hind part of one broken molar and the fore part of the other, he reconstructed a whole molar. This is a distinctive genus evidently of a very primitive ungulate (hoofed) placental herbivore. It was compared with the most primitive ungulates then known from North or South America, but apart from its identity as a genus of ungulates its more precise affinities remain questionable.

In 1967 paleontologists from Montpellier with some Peruvian geologists returned to the site, which is near the shore of a small lake, Laguna Umayo, on the altiplano a short distance west of the great Lake Titicaca. They dug out a large amount of the bed in which *Perutherium* had been found, and this was screen washed. (This technique, now used for collecting scarce or very small fossils involves softening the matrix with water or some other fluid, sifting it through a screen with small holes, and then picking out the fossils left on the screen.) The result was a fragment believed to be a scrap from an upper molar of *Perutherium* and parts of about a dozen smaller teeth, all somewhat broken. These were studied in Montpellier by Bernard Sigé. He found that the small teeth, as far as identifiable at all, were those of opossums, the most abundant North and South American marsupials now and for some tens of millions of years in the past, although in North America they died out for a while and later again spread from South to North America.

Among these tiny, broken, opossum teeth Sigé was able to identify characteristic pieces of upper teeth as closely similar to and at least tentatively assignable to the genus *Alphadon,* first named in 1927 and hitherto known only from the Cretaceous of North America. The Peruvian specimens differ in detail from the several North American species of *Alphadon* that were known by 1971, and Sigé therefore gave them a distinct specific name: *Alphadon austrinum* ("southern *Alphadon*"). There were also some fragments of lower molars that differ from *Alphadon* and are more like the North American genus *Pediomys,* but insufficient to merit definite reference to that genus. The presence of these more or less similar primitive opossums in the late Cretaceous of both North and South America suggests, but is not clear proof, that there was some faunal relationship between the continents at or before that time. Whether they evolved in South or North America and then spread from one contintent to the other by chance or sweepstakes dispersal is not yet wholly clear from the scanty evidence. In any case the now known South American Cretaceous marsupials tie in with the extreme abundance, diversity, and specialization of marsupials that later evolved on the continent but not in North America. It is noteworthy that much better known Mesozoic mammalian faunas of North America were long largely composed of a very different group of mammals known as multituberculates, for the many tubercles or cusps on their molar teeth, and that these are as yet absent from all the early mammalian faunas of South America. (Florentino Ameghino considered some South American fossils to belong to this group, but later study has proved that this was certainly erroneous.)

As regards the Cretaceous mammals from Laguna Umayo, these are so scanty that the comparison with North America may have been greatly biased by chances of collecting. Now, however, there are two more finds made quite recently and some time after those at Laguna Umayo. The first of these new discoveries was made in Bolivia by a party including a North American, several Bolivian geologists, and several French paleontologists. Although politically separated, the high uplifted former trough (now altiplano) extends into Bolivia from Peru. Rocks nearly or quite equivalent in age to those in which the Mesozoic mammals of Laguna Umayo were found also occur near Tiupampa in south central Bolivia, on the eastern slope of the altiplano, nearly sixty miles southeast of Cochabamba and not much farther from more southern Sucre, two small cities easier than Tiupampa to locate on usually available maps.

The expedition was largely supported by the (North American) National Geographic Society through a grant to Larry G. Marshall (currently connected with the University of Arizona). Other personnel and logistic supports were provided by the organizations briefly designated as IBBA (a Bolivian institute for studies of high altitude biology), ORSTOM (a French scientific mission), and the EPHE (a school of advanced studies in Montpellier). Marshall found a mammalian lower jaw with all of it molars, and the party then excavated a large amount of matrix from the fossil-bearing bed. This was shipped to France for screen washing. The screening is being done mostly in Montpellier, but the specimens found will be kept and catalogued in the Parisian Institute of Paleontology of the National Museum of Natural History. I mention this as an example of the nature and extent of international cooperation involved.

As I write this, the results of that labor have not yet been published, but a copy of the manuscript by Larry Marshall, Christian de Muizon (Institute of Paleontology in Paris), and Bernard Sigé (Laboratory of Paleontology in Montpellier) has been made available to me. Eight mammalian specimens were in hand when this was written, all but two of them of the same species, well represented by the jaw with four teeth first found. The specimen is made the type of a species that will be named *Roberthoffstetteria nationalgeographica*. This will not become a valid name for genus or species until published with a diagnosis, which will probably be done before this book appears. The generic name is inspired by Florentino Ameghino's giving such combined names to various early genera, mostly Casamayoran in age (see Chapter 5 here). The even odder specific name is inspired by the fact that a grant from the National Geographic Society made this expedition, and hence this discovery, possible. The other two specimens mentioned as not of this species are said probably also to be opossumlike marsupials but are inadequate for close classification or naming.

The third, most recent, but surely not last Cretaceous mammal to be found in South America was discovered by a young man, Marcelo Rougier, who was with a party led by José Bonaparte. As noted earlier in this chapter Bonaparte, chief of the department of vertebrate paleontology in the Buenos Aires museum, has been mainly concerned with Mesozoic reptiles. While collecting dinosaurs especially in the late Mesozoic (upper Cretaceous rocks) he has constantly had in mind discovery of mammals in those same beds, and on this occasion he was rewarded.

Bonaparte has kindly sent me a copy of the description of this find not yet in press as I write this. The manuscript is by Bonaparte and Miguel Fernando Soria (h), also mentioned earlier in this chapter. The locality is 100 meters west of Cerro Cuadrado ("square hill") in the northern section of the Estancia los Alamitos ("little poplar ranch"), in the southwestern part of the province of Río Negro in northern Patagonia, Argentina. As the specimen was in the same beds as typical late Cretaceous dinosaurs (hadrosaurs) there is no doubt about its age. It is a very small, well-preserved crown of a last upper molar tooth. It is definitely not a marsupial but a placental mammal and apparently a very primitive ungulate. The name proposed for it in manuscript, which will probably be published and authenticated before this book appears, is the new genus *Mesungulatum*, indicating an ungulate in a stage of transition, with the specific name *M. houssayi*. This is in memory of the late Bernardo Houssay, a famous physiologist. He had no special interest in paleontology, and the connection with this find is that he was an organizer of CONICET, an Argentine foundation for the support of science, and that CONICET gave some support to the expedition by which this species was found.

There are quite a few other paleontologists still concerned with South American fossil mammals, but with apologies to those not mentioned here I must draw this chapter, and with it this book, to a close. Throughout I have tried to center discussion on what seem to me personally the most interesting and the most important people and events. There may well be others—both among events and among people—equally noteworthy.

Notes

The brief survey of recent research in this chapter is based on personal acquaintance and correspondence with most, but not quite all, of the people named; on some news reports of their activities; on practically all of their publications; and, as noted, on some relevant manuscripts.

Index

About the Author

George Gaylord Simpson was born on 16 June 1902. His life span thus overlaps many of the paleontologists discussed in this book. His own contribution to the sum of knowledge about fossil mammals of South America has been of signal importance. In addition to many technical studies, he has published several popular books bearing on the subject. The earliest was an account of his first expedition to South America, titled *Attending Marvels: A Patagonian Journal* and published in 1934; it was republished by the University of Chicago Press in 1982. His autobiography, *Concession to the Improbable: An Unconventional Autobiography,* published by Yale University Press in 1978, tells about further field work and studies in Argentina, Brazil, and Venezuela. His survey of mammalian evolution in South America, *Splendid Isolation: The Curious History of South American Mammals,* was published by Yale University Press in a hardcover edition in 1980, followed by a paperback in 1983.